BIRDS and ANIMALS of AUSTRALIA

Rigby Limited • Adelaide • Sydney • Melbourne • Brisbane • Perth
First Australian edition published 1972
French edition copyright © Editions Lito-Paris
English version of text copyright © Rigby Limited
National Library of Australia Card Number & ISBN 0 85179 425 4
All rights reserved
Monophoto-typeset in Australia by Modgraphic Pty Ltd, Adelaide
Printed in Hong Kong

BIRDS and ANIMALS of AUSTRALIA

AND ITS NEIGHBOURS

TEXT BY ROBY

ILLUSTRATIONS BY ROBERT DALLET

EDITED BY J. R. CASLEY-SMITH D.Sc., D.Phil., M.B., B.S., F.R.M.S., F.R.S.M.

RIGBY **OPAL** BOOKS

AUSTRALIA

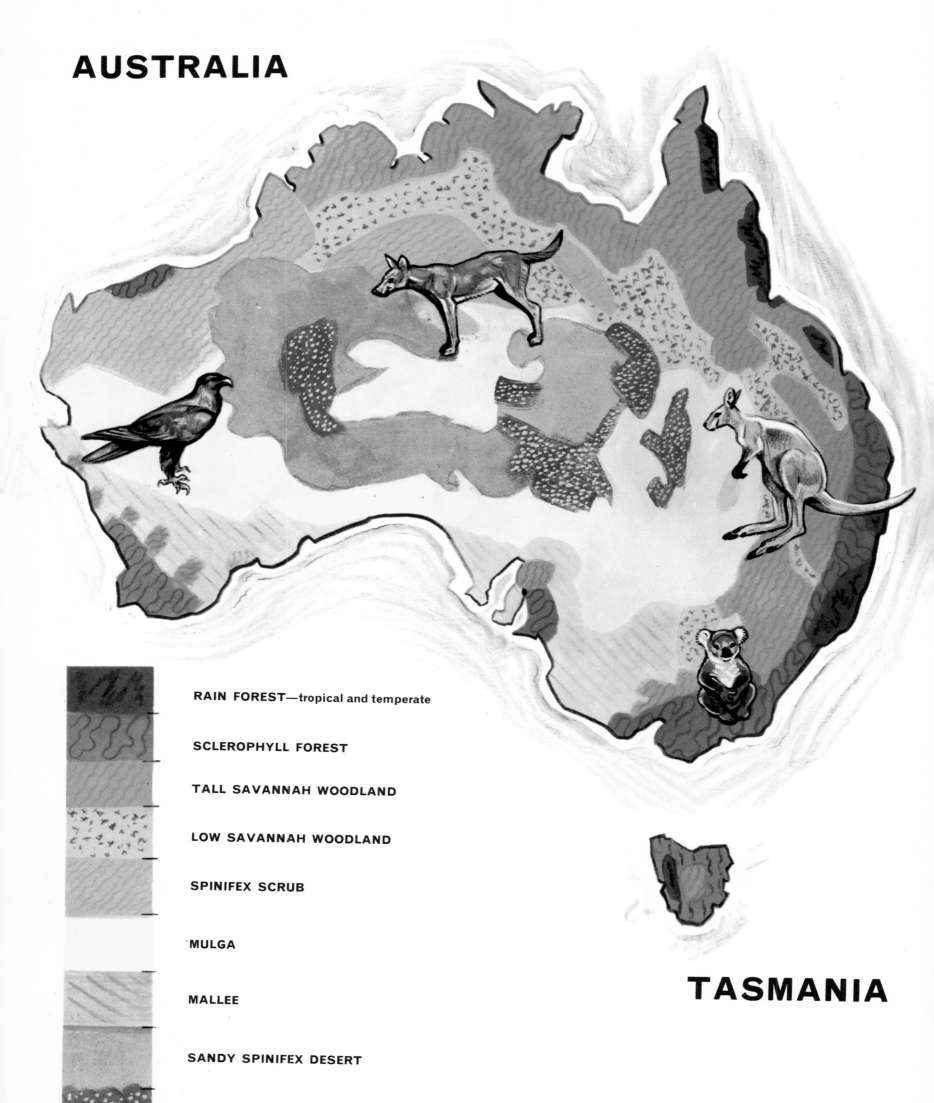

RAIN FOREST—tropical and temperate

SCLEROPHYLL FOREST

TALL SAVANNAH WOODLAND

LOW SAVANNAH WOODLAND

SPINIFEX SCRUB

MULGA

MALLEE

SANDY SPINIFEX DESERT

GIBBER DESERT

TASMANIA

PREFACE

BUCKINGHAM PALACE

The tremendous increase in the human population throughout the world, together with the extent of human technological development, has created a critical situation for wild animals everywhere.

Australia is no exception but, fortunately, there is a growing concern for Australia's wild animals and there is every hope that efforts to protect them from destruction or exploitation will be successful. This is just as well because Australia has a remarkable and unique collection of fauna as a result of its long separation from any other land mass. There have been some catastrophic introductions of exotic animals, the rabbit in particular, but on the whole most of the original inhabitants of the Australian continent and neighbouring islands are still surviving.

If they are to continue to survive, it is vitally important that more people should know more about these strange and primitive animals. It is not enough that professors and scientists should know about them. It is not enough that a kangaroo should be the national emblem or that koala bears are reproduced as cuddly toys. For real safety, these animals must be familiar to children in school and appreciated by adults in all walks of life.

I hope this book will help many people to know and to appreciate the fascinating fauna of Australia.

Duke of Edinburgh

1. LONG-BEAKED ECHIDNA
(Zaglossus brijni)

2. BARTON'S LONG-BEAKED ECHIDNA
(Zaglossus bartoni)

3. SPINY ANT-EATER
(Tachyglossus aculeatus)

The Echidnas comprise two genuses spread over the continent of Australia, Tasmania and New Guinea. The principal genus, which has straight beaks, contains one species (*Tachyglossus aculeatus*). It lives in southern Australia and Tasmania. It is without doubt the best known of this type of monotreme.

The second genus, which has two species, contains *Zaglossus brijni,* or the Long-beaked Echidna, the "Echidna with three digits", or the Proechidna. It lives only in New Guinea and the island of Salwati. You can see this Echidna in the first picture on page 8. It is characterised by its large size (40 centimetres long), long strong legs, and a long beak, sometimes obviously curved. (The drawing seems to show it with a short beak, but this is because its head is turned). The spines disappear with age so that old *Zaglossi* are practically naked. The other species (*Z. bartoni*) is even larger.

The other species of Echidnas is somewhat smaller and less agile. The animals retain their numerous, closely-packed spines, which cause them to be called Porcupine or Spiny Ant-eater. Their Aboriginal names are Gombian or Ningan, as recorded at the end of the eighteenth century by explorers, who saw these hitherto unknown animals in New South Wales and Tasmania. They noticed that the Tasmanian Echidnas were covered with dense fur as well as with spines, because of the harsher climate, and that their very short beaks were straight, like those of the Echidnas of southern Australia. We shall now consider the common characteristics of the three species.

The most astonishing thing about these animals is their famous, pointed beak, in which is their mouth. In fact, it is a beak without really being one, because it does not open in two halves like those of birds, but has a very narrow opening (just below their nostrils), which is actually their mouth. Through this passes their very long, narrow, sticky tongue. This is used by the Echidnas to gather the ants and termites on which they live. One can see this tongue in use in figure 1 on page 10. It is interesting that the Echidna has evolved the same method of feeding as the American Ant-eaters, the African Orycteropus and the different Pangolins of Africa and Asia. They all have the same kind of tongues which shoot out and stick to their prey. The inside of the Echidna's mouth acts like a grater. The tongue has cornified tooth-like projections and the palate has strong, hard crests. With these the insects are broken and the particles of dirt crumbled. This pre-grinding of the food before it reaches the stomach is rather like what happens in the gizzards of birds.

The second peculiarity of Echidnas is their extraordinary burrowing ability. One has only to look at their feet to see how this is possible. Both those with 5 digits (toes), or with 3, have these digits very strongly developed, with massive claws, just right for digging. This is the only effective defence of an Echidna in difficulty—he burrows into the ground and leaves his enemy looking at his erect, pointed spines.

The third peculiarity of the Echidnas is that their hind-feet have enormous claws. These are curved towards the back and the animal supports and anchors itself upon them. One of these claws, the second, seems to reach to the skin of the animal when it scratches itself through the spines. This helps it get rid of parasites, exactly as the Hedgehogs of Europe do.

The fourth characteristic of the Echidnas is that they lay eggs like birds, but suckle their young like mammals. They carry them in a pouch, which is on the ventral (underneath) surface of the females, as in the marsupials. This is shown in the illustration by Robert Dallet on page 10. In figure 2 the egg is shown in its normal size, in figure 3 it is shown much enlarged, while in figure 4 the young Echidna is seen leaving the mother's pouch, which is also called the incubatorium. The body of the Echidna is very supple and she places the egg, as soon as it is laid, in the pouch. Here the young hatches. While it is in the pouch, its spines remain soft, in order not to prick its mother. It does not suckle from a true nipple, but like the young Platypuses, from a slit called the pilifery crypt, which is filled by the milk which runs down the hairs contained in it. When the little Echidna has left the pouch, it is hidden in the bushes and is often visited by its mother. It lives only on insects.

Unfortunately, we know very little about the details of the lives of these strange animals. They have many, very clever adaptations, which are seldom realised from their appearance. They must be well protected so that they do not become extinct.

One can find wild Echidnas in many regions of Australia, New Guinea and Tasmania. One sees frequent signs of their excavations on ant-hills in Queensland, New South Wales, Victoria, South Australia, and Western Australia. The Aborigines like to eat them in the same way that Europeans like roast Hedgehog, but the Echidnas are strictly protected to discourage this. Otherwise, this inoffensive animal, which is a great killer of insects (including Termites, or "White-ants") is left alone by farmers. This is unfortunately not the case with many other Australian species, as will be found in the succeeding pages.

Another favourable factor for the conservation of the Echidna is that they thrive in captivity. They eat milk mixed with the yolks of eggs and finely-minced meat. Some animals have lived in captivity for 30 years.

1. **TONGUE OF THE ECHIDNA**

2. **EGG OF THE ECHIDNA**
   (Natural size)

3. **BABY ECHIDNA
   IN THE EGG**
   (Much enlarged)

4. **ECHIDNA AND HER BABY**
   *(Which is coming out
   of her ventral pouch)*

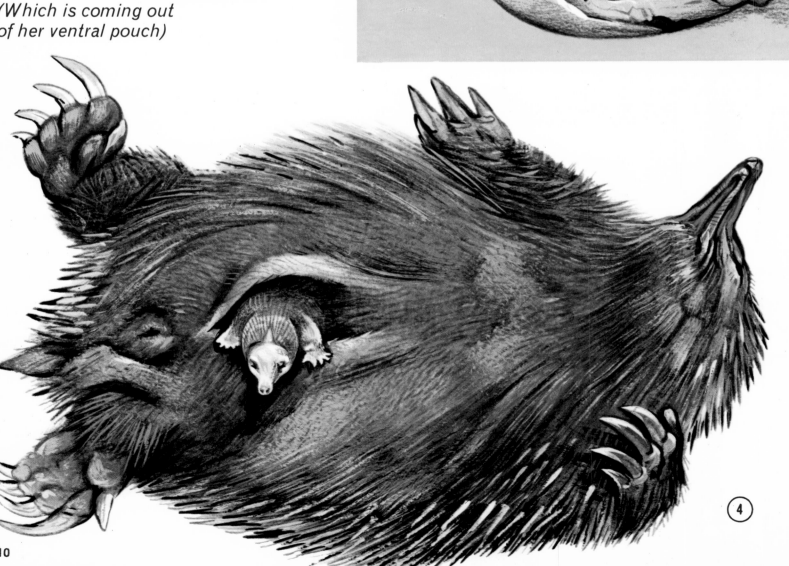

THE DUCK-BILLED PLATYPUS
(Ornithorynchus anatinus)

These animals were first discovered in 1798 in New South Wales. They are rarely seen and were on the road to extinction, but they are now known throughout the world and are famous for their many unique features. Since their discovery, Zoologists have almost continually been able to find new features of interest about the Platypus and each feature seems more astonishing than the one before.

The Platypus and the Echidna, comprise the only representatives of the class of animals called the Monotremes. Although they both have beaks, the beak of the Platypus, unlike that of the Echidna is flattened and opens into two halves. It is more like the mouth of the reptiles than the beak of birds. The Monotremes remain the closest link to the ancient reptilian parents of the mammals. For example, they have only a single ventral orifice; i.e. the urinary and alimentary excreta are ejected through the same opening. This occurs in the reptiles, but not in the mammals, which have two openings. Again, their internal temperature is much lower than that of the mammals and varies much more with the temperature of the surroundings, as occurs in the reptiles. Finally, they lay eggs and their eggs are contained within a soft, leathery shell, like those of snakes and lizards.

There is only a single species of Platypus and it is impossible to confuse this animal with any other. It is 65 centimetres or so long, including the tail. This tail is quite flat, thick, pointed and covered with a brown fur, as is the whole body of the animal. The paws are short, with claws, and completely webbed, as you may see from the illustrations on page 12. The hind-feet are less developed than the fore-feet, but in the male they carry

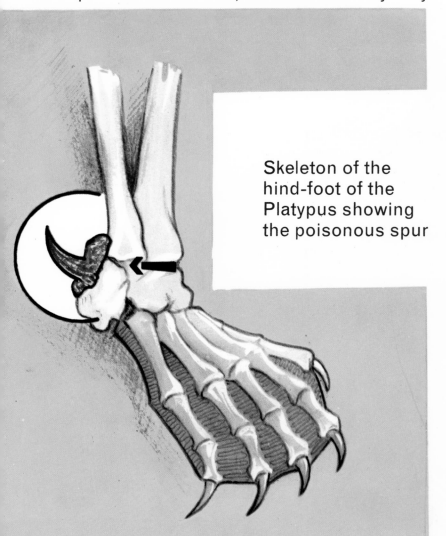

Skeleton of the hind-foot of the Platypus showing the poisonous spur

a poisonous spur, which is fed by a special gland. The poison is quite violent; it can kill little animals and is extremely painful to man. So, apart from certain species of Shrew-mice, it seems that the Platypus is the only mammal in the world with a poison. The illustration on this page shows a paw with the poisonous spur. The artist has shown only the skeleton so that one can see with greater clarity the precise details of this astonishing peculiarity of the Platypus.

In figure 2 on page 10 we can see the egg at natural size. In figure 3 we can see a baby Platypus greatly enlarged. As in the case of the Australian marsupials, the reproduction of the Platypus is not simple. The mother first chases the male Platypus some distance away. Then, once she is settled in a crevice in the bank of a river or lake, she lays from 1 to 3 eggs, which measure between 13 and 16 millimetres in length. These soft, oval-shaped eggs rest one against the other at the bottom of the nest made of moist leaves. The moisture is necessary for their hatching. This occurs after only 12 days, when the baby Platypuses pierce the egg-shell with a special tooth. This is attached to the end of their muzzle and falls off after use, like that produced by certain lizards.

For some weeks the mother looks after her young. She does not have breasts and nipples to give them milk but, like the baby Echidnas, they suckle from hairs on the ventral surface of the mother. The milk is exuded by special ducts at the bottom of a cleft and then runs down these hairs into the mouth of the babies. The eyes of the baby Platypuses open at the end of 11 weeks. They can then eat worms, molluscs and little crustaceans, and they waste no time in starting to chase them for themselves. Their beak is armed on the inside with many cornified ridges, which retain any prey that is sucked in from the water or from the silt at the bottom of the pond.

Platypuses are good swimmers. They pass their time in excavating the bed of their river and its banks, searching for food. These somewhat clumsy animals used to be numerous before the coming of European civilisation (and before the Rabbit). They now exist in Australia only in very localised areas. In other times they were also numerous in Tasmania.

In addition to those Platypuses living in their normal biological habitat, others may be seen in certain zoological gardens in Australia, but there is little hope of seeing them in other countries. This is because the animals are so precious that they are rarely exported and because it is quite a problem to feed them. There have been many good photographic studies of the Platypus made, showing them swimming through large aquariums. These show the manner in which they swim and the way in which they chase their prey. From these films the drawing on the next page has been made to give an impression of what they look like when they are swimming.

All the illustrations contained in this volume have been made with exceptional care, with the result that on the pages of this book we see as exact a representation as possible of each animal pictured.

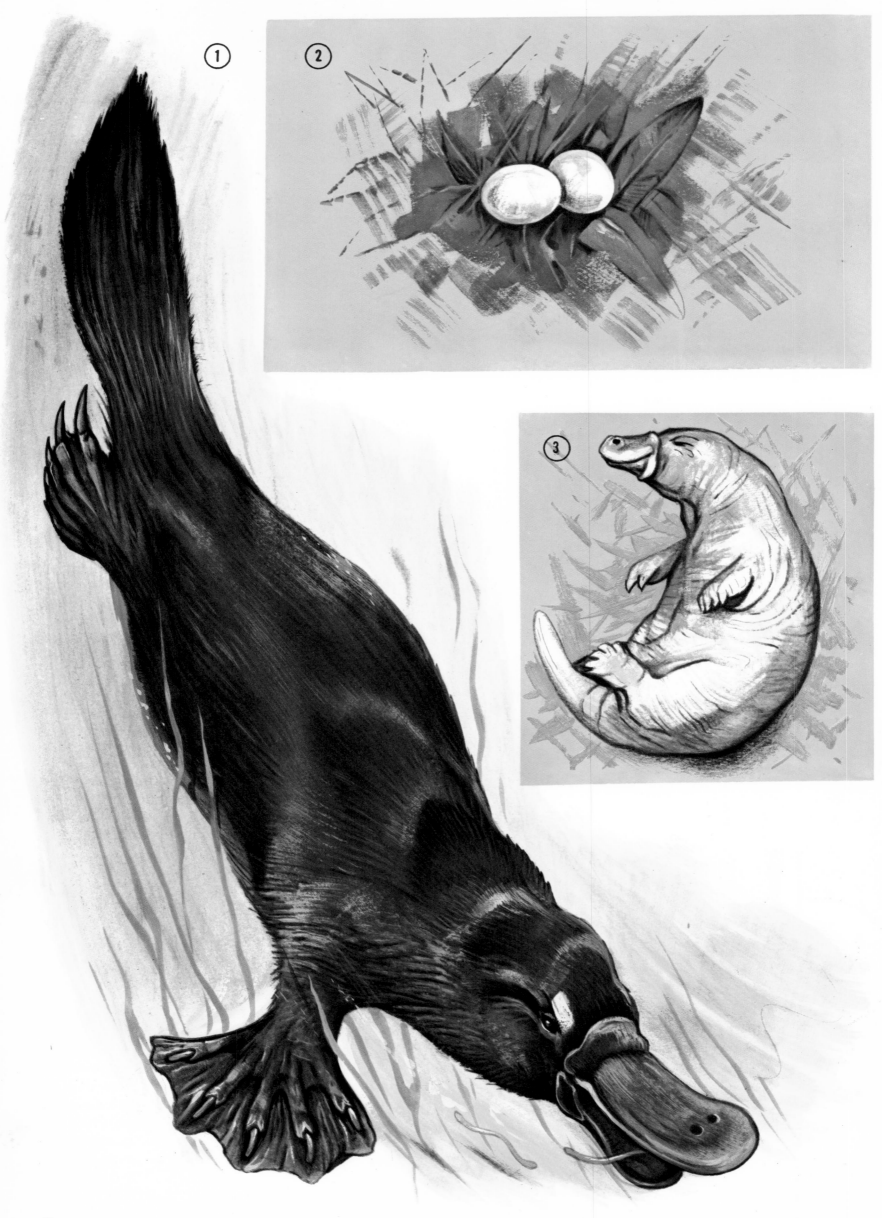

1. PLATYPUS

2. PLATYPUS EGGS
(Natural Size)

3. BABY PLATYPUS
(Enlarged)

THE MARSUPIAL MOLE
(Notoryctes typhlops)

This rather shapeless yellow thing is a mole, but it is a Marsupial Mole—unique to Australia. It leads a life which is almost identical with the life lived by other moles throughout the world. However, it is a marsupial.

These, instead of having well-developed young after a long gestation period, produce very minute offspring, which they care for after their birth in a ventral pouch. Here they complete their development, before they can live freely in the world by themselves.

In Australia the marsupials take the place which is occupied by mammals in the rest of the world, (the sole exception being the Dingo). Thus there are marsupial herbivores, rodents, carnivores and insectivores. These correspond often in their appearance, or at least in their way of life, with the mammalian herbivores, rodents, carnivores and insectivores in the rest of the world. This is a magnificent example of what Zoologists call the Convergence of Forms. By this we mean that the same types of animals occupy the same ecological niches or "places in Nature". The Australian Moles and the European Moles have no common ancestry and yet they have the same mode of life, which has produced almost identical forms. The only point of difference is that the former are marsupials with a pouch, while the latter are mammals and possess a placenta; (i.e. a special organ that nourishes the baby inside the mother's uterus).

The Marsupial Mole was discovered in 1888 by accident and was then described and named in the following year. It created enormous interest among scientists. It has been studied with much enthusiasm, but it has finally become quite rare. Now it exists only in very dry and desert regions, and one wonders how in these areas it can possibly find the worms which form the basis of its food.

The Marsupial Mole is about 15 centimetres long and its fur, which is yellow or white, is thick and short. The hairs are set perpendicularly in the skin, so that the animal can advance or retreat in its subterranean galleries, without ever finding that its hair has been rubbed up the wrong way. It has no traces of eyes nor of external ears; even its tail is minute. The cornified skin of its muzzle is lubricated by a duct which carries tears from its lachrymal glands and which discharges near the tip of its nose. This prevents its muzzle from becoming encrusted with dirt. Both its front and back feet have 5 digits, but the front ones carry much larger claws. On the third and fourth digits the claws are particularly well developed and allow the animal to dig rapidly and vigorously into the soil. It digs as well at a depth of 60 centimetres as it does on the surface, and the light-coloured soil thrown up allows one to trace its galleries. From time to time it seems to want to come out into the open air. When this happens, one sees a sinuous triple-track which has a deep hollow in the centre where the body of the mole went, and two smaller hollows on each side formed by its paws.

It has an enormous appetite, like the other "true" Moles. It seems capable of eating not only worms, but also all sorts of insect larvae and perhaps even certain roots. Actually one knows very few details about this rare animal. It is impossible to keep it alive in captivity and it is rarely discovered in its native habitat. It seems that there is another variety of Marsupial Mole, even more rare, which lives on the western boundary of Australia next to the Indian Ocean. We know very little about this animal, except that its muzzle is a little shorter than that of the normal Marsupial Mole.

We can only guess about the details of the reproduction of these animals. We know neither the length of their gestation time, nor the length of time which the young spend in their pouches. The pouches open towards the rear, as is the rule with all marsupials which do not stand upright, or sit, or climb trees.

The ignorance which we have to admit about the way of life of the Marsupial Mole is unfortunately going to be repeated with respect to many other animals during the course of this book. The animals of Australia, in particular the marsupials, are some of the strangest, most baffling, and least-studied animals in the world. Our information about the animals we do know of is far from complete; there may even be animals which we do not know about, such as the famous Striped Tiger Cat of Cape York in Queensland. There is considerable argument about whether such an animal really exists.

1. THE SAND-LOVING MARSUPIAL MOUSE
(Sminthopsis psammophila)

2. THE WHITE-BELLIED POUCHED MOUSE
(Antechinus flavipes leucogaster)

3. THE FAT-TAILED MARSUPIAL MOUSE
(Sminthopsis crassicaudata)

4. CANNING'S LITTLE PUPPY-DOG or MULGARA
(Dasycercus cristicauda)

5. THE JERBOA MARSUPIAL MOUSE
(Antechinomys spenceri)

This group of rodents corresponds very well with the mice, rats and gerboas of other continents. There are many diverse types, which are grouped under the general heading of Dasyurids. Some are quite large and carnivorous, others are very small and insectivorous, but all are nocturnal. It is among the Dasyurids that one finds the Phascogales, which we will meet in the following pages. Many of these animals have two common names. The most ancient is that given by the Aborigines, to which has been added the name given by the first European colonists. Without reference to the true Zoological nature of the animals, these colonists gave them names from the European animals with which they were familiar: mice, rats, gerboas, rabbits, squirrels, cats, wolves, badgers. These have been applied to the animals found in Australia, although they have no affinity whatsoever with the European animals, which they may perhaps resemble. This has often complicated matters.

In general the animals on the next two pages live off insects of all sorts.

They have enormous appetites and build nests in holes in the ground or in cracks in trees. One can find a few of them anywhere in Australia, in the bush, in the forest, and even in the most desert regions. Certainly they have many of the characteristics of marsupials, but not all of them have the famous ventral pouch—the marsupium, which is normally so characteristic of these animals. In many of the species the marsupium is reduced to a simple fold of skin on the ventral surface, behind which are hidden the nipples.

This is another example of the phenomenon of Convergence of Forms, which we found with the Marsupial Mole. There is a particularly good example, which concerns the Jerboa *Antechinomys*. Unlike the true Gerboas, which are rodents and live in Africa and Asia, this is an insectivore with some carnivorous tendencies. However, its size, its general body form, its long, balancing tail, its immense hind-feet and its minute fore-paws are exactly like those of the Gerboas in the ancient world. Once again the same sort of country and the same way of life produce the same effect—an animal which has almost the identical form to that of quite another animal, which lives in what is called the same ecological niche. In this instance the two animals are not only not of the same Family, but even of the same Order.

The other Dasyurids, which have the names Marsupial Mice and Marsupial Rats, do not really have a very great resemblance to the mice and rats of Europe. On the next page one can see the Phascogales, which have the same way of life as do the Weasels, but which hardly resemble them at all.

All these animals have enormous, brilliant eyes, which are useful in their nocturnal living. They have beautiful fur, which is always extremely clean. They have most amusing attitudes, with quick abrupt movements, and frequently sit with their very tiny "hands" holding things just under their mouths. They are the little imps of the Australian night.

⑤

1. THE BROAD-FOOTED PHASCOGALE
(Antechinus)

2. THE PIG-FOOTED BANDICOOT
(Chaeropus ecaudatus)

3. THE BRUSH-TAILED PHASCOGALE
(Phascogale tapoatafa)

4. BYRNE'S POUCHED MOUSE
(Dasyuroides byrnei)

Following on from the last page, here are some more Dasyurids. We have two Phascogales, a Bandicoot and a Marsupial Mouse. These live in the desert and the tails of the latter two are covered with long, black hair.

Phascogales and Bandicoots are animals which are actually very closely related, although the Broad-footed Phascogale seems to be much more closely related to the Marsupial Rats. Even more interesting is the Pig-footed Bandicoot which we show at the top of page 17. This is so named because it has only 2 digits on its front feet, as do the pigs, but the form of this animal is more like that of a little Antelope. And indeed it can run very quickly. It is only 25 centimetres long, not counting its tail, which is about half as long as its body.

There is a very funny story about this animal. It was discovered in 1836 by the explorer, Mitchell. The first specimen of this animal captured had accidentally lost its tail and so Mitchell called it *Chaeropus ecaudatus*, which means "*Chaeropus* without a tail". It has, of course, since been discovered that this animal has a perfectly good tail. Unfortunately, however, the first Zoological description, even if ridiculous, has priority. So for numerous Zoologists the Pig-footed Bandicoot is always called *Chaeropus ecaudatus*, although others have re-baptised it *Chaeropus castanautis*.

As well as having strange front feet, *Chaeropus* is interesting because of its way of life. When living in the wild, it eats only vegetation. Thus it is halfway between the Phascogales, which are often carnivores, and the large marsupials such as Wallabies and Kangaroos, which are herbivores. These will be considered later. On its ventral surface *Chaeropus* possesses a marsupial pouch which opens towards the back. This contains 8 small breasts, but there are never more than 2 young ones in the pouch. The animal makes a very well-sheltered nest, containing much dried vegetation. This is usually in a hole in the ground, but can equally well be in a crack in a tree.

In 1900, *Chaeropus* was considered extremely rare and some people now think it has completely disappeared. It would be wise not to commit oneself here, however, because happy surprises have occurred from time to time for Zoologists who are interested in rare animals. One thing is certain and that is that the introduction of the Fox to Australia has completely disrupted the natural equilibrium of the native animals. Foxes have killed enormous numbers of animals which have been unable to escape from them and which are unable to reproduce themselves rapidly—as is the case of the poor Pig-footed Bandicoot.

The Brush-tailed Phascogale is slightly smaller, it is more highly specialized, and it is a carnivore which climbs trees. One can compare them with the weasels and stoats of Europe, Asia, Africa and North America, which they resemble from many points of view. As for the little Byrne's Pouched-mouse, at the bottom of page 17 this is one of the numerous desert "rats", which have been called this in spite of the fact they do not act like rodents at all.

Once more we find a Convergence of Forms by different animals (in spite of their having no relationship) because their ways of life and the ecological niches which they occupy are very similar. This phenomenon is not restricted to the animals of Australia, but it is here that one finds some of the best examples. It is one more reason why we should hope that the Government of Australia will take effective measures to protect these unusual animals.

1. THE RABBIT-EARED BANDICOOT
(Macrotis lagotis)

2. THE BARRED BANDICOOT
(Perameles fasciata)

3. THE STRIPED BANDICOOT
(Perameles gunnii)

4. THE SHORT-NOSED BANDICOOT
(Thylacis isoodon)

5. THE LONG-NOSED BANDICOOT
(Perameles nasuta)

The name "bandicoot", which was originally taken from the Hindu for "pig-rat", has been applied to a group of marsupials with medium-length tails, which have something of the life-habits of the *Lagomorphs* (i.e. Hares and Rabbits), which are found in the rest of the world. They are often called by the general name Perameles or Peramelids, which comes from their scientific name.

We are going to talk mostly about the Rabbit-eared Bandicoot, which has the common name of "Bilby", and which is quite popular. In order that you can see something of the other Peramelids, there are a number of them drawn on page 19. All these animals are characterised by a very beautiful short fur, which is uniform in colour in certain species, but has two colours in others. They have short tails which are sometimes naked and sometimes hairy. Their muzzles and ears are very similar to those of certain types of rats of different species, which have the life-habits of insectivores, carnivores and herbivores. Notice also that their hind-feet are very long, with a single, central claw and two shorter ones on either side. (This characteristic is found also in the great herbivorous marsupials.) The Peramelids do not dig holes, and use their hands only to carry food to their mouth like the Phascolgales in the preceding pages. The Peramelids have seven different genera which are found from one end of Australia to the other. Each genus has many different species in it.

Now for Bilby. This is a pretty, graceful animal, which can move very quickly. He has external ears like a Rabbit and very thick fur of a lovely grey colour. His beautiful, long, fur-covered tail is black on top where it joins his body, and then becomes white. The Aborigines love to eat him and wear his tail. He digs holes and in other ways resembles the Rabbit.

Bilby is usually nocturnal. He eats all sorts of small prey, insect larvae, molluscs and numerous little mammals which he happily kills. Actually he has very good teeth, both canines and molars, His reproduction is not very rapid—only two young per year. Because he is pursued by the Aborigines (both for meat and for his tail), because of competition by the imported Rabbit, and because of the advance of civilization, Bilby is disappearing more and more rapidly. He lives only in semi-desert regions in the south-west of Western Australia. While he was widespread in the inland in earlier days, even at the beginning of this century he was already becoming rare.

Finally, before leaving Bilby and the other Peramelids, we should say that these animals have a method of reproduction which has some features in common with those of the placental mammals. Their young are born relatively much larger than those of the other species of marsupials. However, they still grow considerably in the maternal marsupial pouch.

THE TIGER CAT
(Dasyurus maculatus)

These animals take the place of cats in Australia. The largest is the one which Robert Dallet has drawn above—the Tiger Cat. The existence of a very large Striped Tiger Cat in Cape York Peninsula remains to be proved. The spotted Tiger Cat was discovered in 1789 by some of Governor Phillips' men. It is a beautiful animal about 60 centimetres long, with a tail which is roughly the same length. It has a tawny coat with some white irregular spots on it. While it has the strong canine teeth of a hunter, it does not have the retractile claws of the true cats. On the contrary, the bottom of its feet are adhesive, which fact permits it to climb very well. It spends most of its time in the trees.

It has 4 to 6 babies a year, and these are born in June. The marsupial pouch is formed only by a few folds of skin on its ventral surface. The babies are therefore placed in the bottom of the parents' lair, where they stay and are reared.

The nocturnal Tiger Cat hunts with great perseverance, cunning, patience and much force. It will even stand up to a dog and can kill a wallaby or a possum. It is very fond of birds and our artist shows one with a parrot in its jaws.

While the area in which the Tiger Cat lives is diminishing, one can still find it in Tasmania and in the south-east quarter of Australia. However, it is an animal which rarely shows itself. One is more likely to pass it by without being aware of its presence, unless one surprises it in pursuit of its prey. Nowadays it has become even warier, because it has learnt to its cost what happens if it is seen by neighbouring farmers.

THE EASTERN NATIVE-CAT
(Dasyurus viverrinus)

This is similar to the Tiger Cat but rather smaller, with shorter legs. It exists on the mainland of Australia and in Tasmania. The animal is shown on page 21 and is sometimes called the Australian Cat or the Eastern Indigenous Cat, because it is in this region that it is found, as is also the case with the Tiger Cat. It also feeds on all sorts of living prey—rats and mice, little birds, young rabbits, small reptiles, insects and even domestic fowls when they are in its territory.

There is another animal which used to be extremely common in Australia, but which is now quite rare. This is the Quoll (*Dasyurus quoll*), which is very similar to the Eastern Native cat, except that it is more marked by pale spots on its tawny-grey coat. In the Aboriginal legends, it was held that the pale spots were wounds which the animals received in the course of their tribal warfare, similar to the warfare of the Aboriginal tribes. Thus these animals, in common with their homologous animals in the rest of the world, have played, and indeed still play, an important role in the local folklore.

It is evident that these wild Australian cats are not viewed in any better light than are the wild cats in the rest of the world. They are shot, poisoned, trapped and chased by specially trained dogs, so that their numbers are greatly reduced. In addition, they face severe competition from imported domestic cats. They have also inherited all the hatred which for thousands of years people in many parts of Europe have felt towards these minor feline predators. It is therefore a very important task to convince people that these carnivores deserve complete protection, if only in reserves.

THE NUMBAT OR BANDED ANT-EATER
(Myrmecobius fasciatus)

The Numbat and the Echidna are to Australia what the other ant-eaters are to Africa, Asia and South America. Again we find the same Convergence of Forms. Unfortunately the Australian Numbat is becoming extinct at a very rapid rate. It was identified in 1836 and two varieties have been described at the extremities of the Great Australian Desert. The eastern variety seemed to be extinct, when several specimens were recovered from around Mount Everard in the north-west of South Australia.

The Numbat is about the size of a large rat, but it has a beautiful furry tail, which is as long as its body. Its coat is tawny with a certain amount of red in it, and is striped with large bands, alternatively black and white, down to the beginning of the tail. One type, which is found in eastern Australia, is not banded and is of a uniform red colour. The head is long and flat, and its jaws, although they possess 52 teeth, are quite feeble. Its tongue is cylindrical and is only about 12 centimetres long, but the way in which it captures ants and termites is the same as that used by the other ant-eaters. It does not have very strong nails to dig out the termite nests, of which the inhabitants, together with ants, are all that it eats. It therefore prefers to hunt for termites in the rotten wood which is infested by them.

We do not know whether the Numbat swallows its prey whole or whether it chews it up first. Probably it uses both methods, particularly since its jaws are not joined together as are the jaws of other ant-eaters and it can open its mouth. This is then a facultative ant-eater; i.e. it is just beginning to specialize in eating ants but it can probably eat other food as well. It comes from a branch of the Dasyurids, which are becoming more specialized to this form of life. This is probably because termites abound throughout all of Australia.

The females do not have a marsupial pouch. So, after being born, the 3 or 4 baby Numbats are caught in the fuzzy hair which covers the ventral surface of the mother. They then become attached to her nipples and remain there. Later on we will see that the great marsupials also grow in this fashion, either inside or outside the marsupial pouch. If one tried to detach them from the nipples, one would have to tear their mouths. Thus the little Numbats, which are only about 2 centimetres long, are able to remain attached very firmly to their mother, no matter what she does.

This page:
THE SQUIRREL GLIDER
(Petaurus norfolcensis)

Following pages:
1. THE SPOTTED CUSCUS
   (Phalanger maculatus)
2. THE NORTHERN BRUSH-POSSUM
   (Trichosurus arnhemensis)
3. FLUFFY or YELLOW-BELLIED GLIDER
   (Petaurus australis)
4. THE SUGAR GLIDER
   (in flight)
   (Petaurus breviceps)

On this page we see the Squirrel Glider. On the following pages there are shown several other Australian animals which use the same means of locomotion, which is also known by other animals (particularly the squirrels) elsewhere in the world. It consists of holding out a "patagium", or gliding membrane. This is a fold of skin which is stretched by the four limbs and acts as a wing on which the animal can glide. They sometimes cover distances of more than 100 metres, provided they start from a fairly high point.

The Squirrel Glider is one of a group of animals which in Australia are called Possums. It is important not to confuse these Possums with the Opossums, which are also marsupials, but which are found in America. The first explorers who examined the Australian fauna described these Possums and forgot to put in the "O". Over the years this mistake has been preserved, so that there is no danger of our confusing Possums and Opossums.

But coming back to the Squirrel Glider, it should be realized that it lives almost entirely in trees; i.e. it is an arboreal animal. It eats basically a vegetarian diet, but also eats larvae, insects, eggs and perhaps also, on occasions, small birds. Once more we have another example of this extraordinary Convergence of Forms between the Flying Possum of Australia and the Flying Squirrel of America and northern Asia. They have the same soft fur, the same white belly, the same long hairy tail, the same short limbs which terminate in prehensile hands (i.e. capable of grasping things), the same large, black eyes and the same little ears. Only the colour of the fur and the black lines on the back, in the middle of the face, around the ears and on each edge of the membrane differentiate them. People who know the animal well say that it gives a little cry when it jumps into the air. There are many different varieties of Gliders, which are quite prevalent in Australia.

It is quite certain that the older inventors did not know of the flying marsupials of Australia, but they could quite well have known of the flying mammals of Russia and Siberia. Perhaps Leonardo da Vinci conceived the idea for the first model of a parachute after having talked to someone who knew of these animals. This parachute had a sort of rigid frame, shaped like a cross, to which was attached light material. This was in turn attached to the legs and hands of a man, who was supposed to jump into the air from the top of a tower. All that stopped this early invention from being practical was its small amount of surface-area compared with the weight of a man. One should compare the area of the membrane of these flying animals with their weights. What might not Leonardo da Vinci have done had he had a specimen of one of these animals actually in his hands! He knew that nature always has a lot to teach us.

23

On these two pages are shown four more Australian Possums. In figure 1 there is the Spotted Cuscus (*Phalanger maculatus*), in figure 2 another Possum (*Trichosurus arnhemensis*), in figure 3 the Fluffy or Yellow-bellied Glider (*Petaurus australis*) and in figure 4 the Sugar Glider (*Petaurus breviceps*) in full flight.

The Cuscus lives in trees, but does not fly. Instead he climbs with the aid of a prehensile (grasping) tail. He has an odour reminiscent of musk, like that found in Skunks. He is the biggest of the Possums and his long tail is furry when he is born, but the end-half rapidly becomes naked. It forms a sort of fifth hand and the Cuscus always has it curled round a branch when he sleeps, which is by day. At night he chases all sorts of small, living prey as well as eating shoots and leaves. In New Guinea there is another variety of Cuscus with a lighter coat.

The animal drawn at the bottom of page 24 is a Possum very like the Cuscus. He is called the Northern Brush-Possum and is well-known in other countries. Since the beginning of this century, many millions of pelts have been sent from Australia to Europe and America. This is because the fur of this little animal is rather like a Fox and is very much in demand. It is uniform, sometimes grey, sometimes brown, sometimes white.

The third Possum (which is sometimes called the Petaur or Australian Opossum, thus confusing its name even more), lives around the continent in the coastal regions. It does not fear man and it often enters houses, if it finds a window open, to get food and shelter. It eats fruits, berries and grain. This often earns it the dislike of man because it takes fruit, nuts and grapes from householders. Finally, in some ways this pretty, nocturnal animal reminds one more of a large Dormouse than of a little Fox, less because of its way of life than because of its looks. It is, however, not a rodent.

The Sugar Glider, on page 25, shows peculiarly well the membrane or patagium of the flying possums. One can see very well how this membrane is attached to the hands and feet of the animal and how its beautiful tail acts as the rudder and elevators do in an aeroplane.

24

(1)

1. THE PIGMY GLIDER or FLYING MOUSE
(Acrobates pygmaeus)

2. THE GREATER GLIDER
(Schoinobates volans)

3. THE GREATER GLIDER
(in flight)

Two very characteristic flying Possums are shown on these two pages: actually they are the dwarf and the giant of this group. The first is shown here actual size. The second, which was discovered in 1789, is also called the Black Flying Opossum and is the size of a big rabbit.

The Pigmy Glider on this page shows on its flank the white hair of its patagium, folded up. Its tail has a particular form, being pointed and as long as its body, with radial hairs sticking out on each side. This allows the tail to act as a very good rudder and elevator assembly, which is both supple yet sufficiently rigid for the animal to be able to use it to direct its flight very accurately. It is able to change its course up, down or to either side. This little animal is very light and cannot make long flights, but it is extremely agile and can move from branch to branch very rapidly. It is in fact very difficult to follow its course, particularly as it moves mostly at night. The mother often has four young, which she keeps for a long time in her marsupial pouch, before they are capable of clinging to her back. The adults make a round nest of Eucalyptus leaves. These animals eat termites taken from dead branches, and honey taken from the flowers of the trees in which they live.

The Greater Glider has been shown on the opposite page in two views. The first shows a female clinging to a tree with the young one on her back; the second, a male in full flight. One can see at once that the fur of the three animals is not the same. The male is blue-grey, while the female is a chestnut and brown colour. However, when these animals are flying through the branches, it

is extremely difficult to say what colour they are, since they move so rapidly.

One should notice the rather triangular form of their body, when they spread out their membrane. This is because it is attached to their feet at the rear but only to their elbows at the front. Thus they have something of the appearance of a delta-wing aircraft. The tail is very characteristic. It is long, well-furred, and flat, forming an excellent aid in their flying. The head is also very remarkable with the large, round eyes which are so characteristic of nocturnal animals, and enormous ears which stick out widely from their heads and are fringed with long hairs.

These animals are strictly vegetarian. They eat only the buds of Eucalyptus trees. Many of the Australian animals show decided preference for this tree. The best observations show that the Greater Glider can fly for distances of up to 120 metres. These flights allow them to live indefinitely amongst the trees, without coming to the ground. The great flyers can climb to a high take-off point and then manoeuvre very accurately with their tail and membrane, so that they alight on another tree without touching the ground.

They have only one baby at a time, and while it rests in the marsupial pouch there is no great difficulty in transporting it. As with the other varieties of Possums, a long thin nipple is permanently in place in the mouth of the young. The whole pouch is filled with a viscous solution which acts as a shock absorber for the baby. Everything changes when the young finally leaves the pouch and has to live the same life as its parents, without being capable of their flying performance. The solution we see on the opposite page. Like the little Koala, which we will discuss later, the baby Glider clings tightly to the mother's woolly fur. Thus the young early learns lessons of aviation, and appreciates its good fortune in being able to fly.

If one looks at the drawings by Robert Dallet on these two pages, as well as on some of those preceding and following, one can see that certain of these Possums (either flying or not) have an unusually shaped hand. Instead of having a thumb with 4 opposing digits, they have 2 thumbs with 3 opposing digits. It is as if our index finger had become another thumb. One can see this very distinctly in the hand of the Pigmy Glider on page 26 and in that of the Greater Glider on page 27. This must surely give a much better grasp to these animals when they arrive at their destination and have to grab hold of a branch to support themselves.

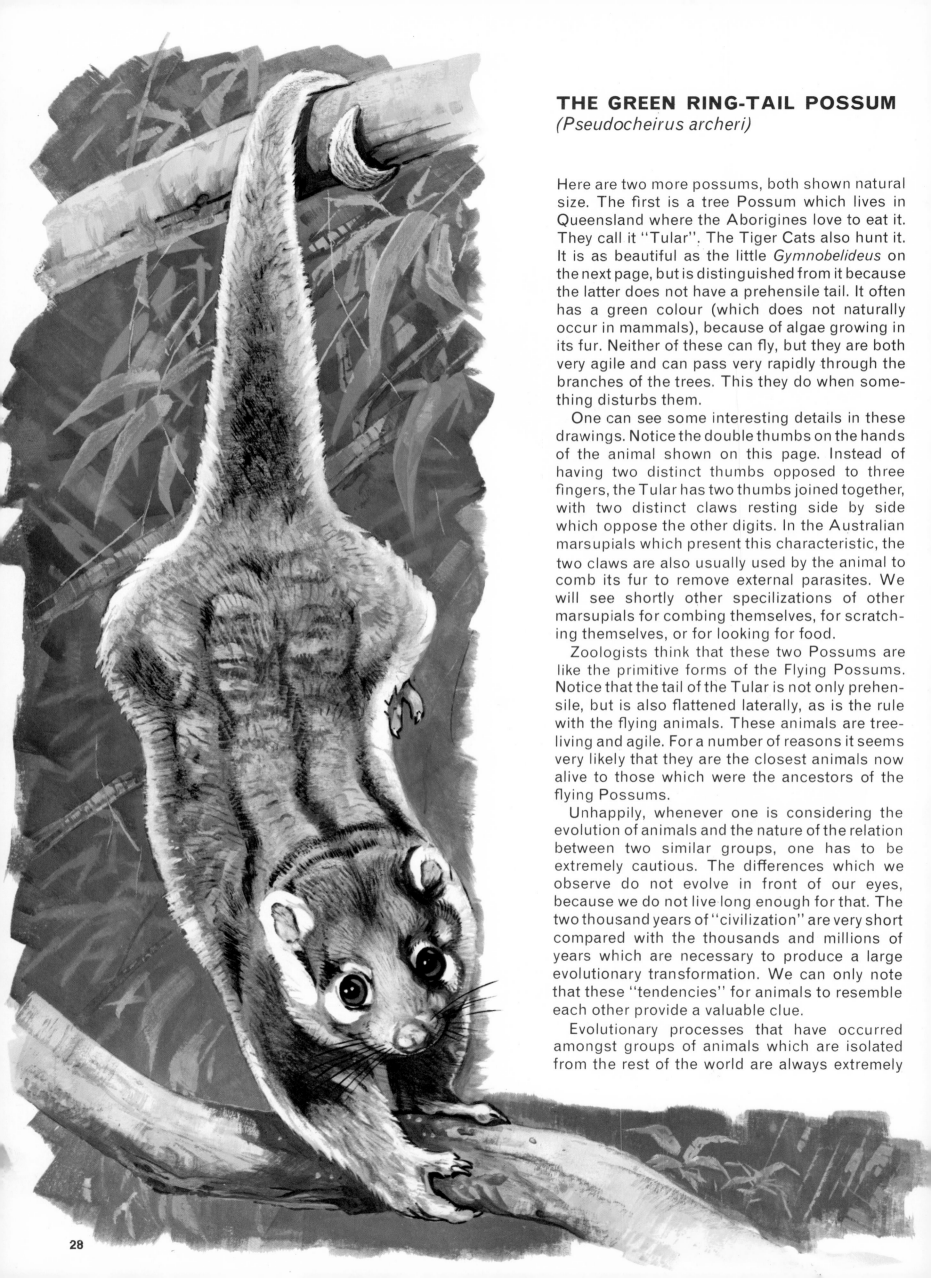

THE GREEN RING-TAIL POSSUM
(Pseudocheirus archeri)

Here are two more possums, both shown natural size. The first is a tree Possum which lives in Queensland where the Aborigines love to eat it. They call it "Tular". The Tiger Cats also hunt it. It is as beautiful as the little *Gymnobelideus* on the next page, but is distinguished from it because the latter does not have a prehensile tail. It often has a green colour (which does not naturally occur in mammals), because of algae growing in its fur. Neither of these can fly, but they are both very agile and can pass very rapidly through the branches of the trees. This they do when something disturbs them.

One can see some interesting details in these drawings. Notice the double thumbs on the hands of the animal shown on this page. Instead of having two distinct thumbs opposed to three fingers, the Tular has two thumbs joined together, with two distinct claws resting side by side which oppose the other digits. In the Australian marsupials which present this characteristic, the two claws are also usually used by the animal to comb its fur to remove external parasites. We will see shortly other specilizations of other marsupials for combing themselves, for scratching themselves, or for looking for food.

Zoologists think that these two Possums are like the primitive forms of the Flying Possums. Notice that the tail of the Tular is not only prehensile, but is also flattened laterally, as is the rule with the flying animals. These animals are tree-living and agile. For a number of reasons it seems very likely that they are the closest animals now alive to those which were the ancestors of the flying Possums.

Unhappily, whenever one is considering the evolution of animals and the nature of the relation between two similar groups, one has to be extremely cautious. The differences which we observe do not evolve in front of our eyes, because we do not live long enough for that. The two thousand years of "civilization" are very short compared with the thousands and millions of years which are necessary to produce a large evolutionary transformation. We can only note that these "tendencies" for animals to resemble each other provide a valuable clue.

Evolutionary processes that have occurred amongst groups of animals which are isolated from the rest of the world are always extremely

LEADBEATER'S POSSUM
(Gymnobelideus leadbeateri)
(natural size)

interesting for Zoologists, because they produce a phenomenon in a closed system, which one could not otherwise observe. This is why the Galapagos Islands (which are 1000 kilometres from the coasts of South America, on the equator, in the middle of the Pacific), form one of the most precious laboratories which exist in the world. It was here that Charles Darwin first had the idea on which he based his theory of the Evolution of the Species. While this theory has since been modified in some details, it was the door which led to all modern Zoology.

It is the same in Australia. This continent has also been isolated for a very long time. Except for birds, the insects, the marine animals and those animals imported by Aborigines and Europeans, (in particular the Dingo), this continent, up to the beginning of the eighteenth century, had kept its special animal fauna isolated from outside contacts.

Then came the Europeans, who brought to Australia animals not previously seen here. These animals had a placental reproduction. Their babies, when born, were much larger and stronger than those of the marsupials. This gave them a considerable advantage over the marsupials, which were by comparison small and feeble. Dogs, cats, rabbits, pigs, sheep and cattle, whenever they were placed, multiplied magnificently, but always at the expense of the local animals. Dogs and cats are certainly responsible for the death of large numbers of relatively defenceless Australian herbivores, which previously had not known any enemies. Even the simple presence of rabbits, with their hundreds of burrows per acre, was sufficient to push out many Platypuses and Echidnas. There is considerable competition between animals which eat the same sort of food. Bilby, the Bandicoot, which has already been mentioned, has already become rare in the Australian plains. The millions of rabbits which have developed there have virtually eliminated him, just as they have eliminated many other herbivores which have been overwhelmed by their numbers.

If one adds to this the pressure of human development and man's ferocious hatred for all things which do not seem to him to be useful, and which he therefore calls "nuisances", one can see what is going to happen.

1. THE MOUNTAIN POSSUM
(Trichosurus caninus)

2. THE TASMANIAN BRUSH-TAILED POSSUM
(Trichosurus vulpecula fuliginosus)

3. THE COMMON RING-TAILED POSSUM
(Pseudocheirus peregrinus)

4. THE BRUSH-TAILED POSSUM
(Trichosurus vulpecula)

The *Trichosuri,* and specially the species *vulpecula,* are spread out through the continent of Australia. They are well-known to man and show varying colours in their fur. There are local varieties, brown, grey and sometimes nearly Albino. Some have a prehensile tail, and one can see here two species which have this peculiarity (figures 1 and 3). The others (figures 2 and 4), have a thick, furry tail, which is rather like the tail of a Fox, as is well shown at the bottom of the page.

All these animals love tender leaves, grains, berries and fruits. They frequently steal fruit from orchards and grapes from vines, and are therefore disliked by farmers and householders. Some also eat a little meat (e.g. dead birds), but they are not carnivores in the proper sense of the word, because they do not hunt animals. They prefer to remain in trees, but sometimes run on the ground and sometimes hide in crevices of rocks. They all have the most beautiful fur and have been much hunted for their pelts, to make coats. This has not made them fear man. In fact they are some of the animals who fear him the least, in spite of their being amongst the animals which he kills the most. They often become tame by themselves and seem to appreciate the tranquil life which is offered to them by a zoo or friendly house.

The European animals which they most resemble are the Fox, the Martin, the Squirrel, the Dormouse and the Rat. Because they are not completely rodents and because of their furry bodies, and the fact that they are furtive and yet cheeky, it is perhaps the Dormouse which they most resemble. One can often hear *Trichosurus* galloping along iron roofs, or fighting amongst themselves in the trees, as do the Dormice. They also eat the same sort of food and have the same way of life as these animals.

In general they live in little groups comprising a single adult male and a certain number of females. These he will have won after noisy, vigorous battles with his rivals. They have one or two babies in July, after a gestation of about two weeks. These mature to become adults seven months afterwards.

1

1. THE SOUTH-WESTERN PIGMY POSSUM
(Cercartetus concinnus)

2. THE STRIPED POSSUM
(Dactylopsila picata)

3. THE EASTERN PIGMY POSSUM
(Cercartetus nanus)

4. THE HONEY POSSUM
(Tarsipes spencerae)

All the animals shown on these two pages are drawn in their natural size. We see the two smallest marsupials to be found in all Australia. In general they are all Possums. This name, which is purely a local way of speaking, has no formal scientific value, since it designates all sorts of small arboreal animals coming from many different species and genuses.

Cercartetus Concinnus and *Cercartetus nanus* are minute flying Possums considered by Zoologists to belong to a particularly old genus. It is thought that from these, as well as from the Pigmy Glider *Acrobates pygmaeus* and also from the Sugar Glider *Petaurus breviceps* (both of which we have already considered), the other forms of flying Possum have been derived. They all live in trees and let themselves fall from the branches to begin their flight. People who do not know them think that they are some sort of pretty mouse, but these are mice who have learnt to fly—as in a fairy-tale.

The *Dactylopsila* is a very beautiful Possum which is quite rare and lives in northern Queensland. It is called the Striped Possum, which is easily explained when one sees the beautiful black-and-white stripes on its flanks and head. We do not know very much about this animal, except that it breaks into the hives of wild bees and eats them and their honey. On the end of its tail, among the black-and-white hairs, it has a sort of curved spine, the use of which we do not know. We know that the same sort of curved spine is found in the end of the tails of Lions, but we cannot explain its presence in either case. You can see that the Striped Possum has one very long finger, the fourth, on each of its hands. This has a pointed nail on its end and is probably used to capture the larvae and insects which it looks for under the bark and in holes in trees, and which constitute its food. The "Aie-Aie" of Madagascar has a similar peculiarity and uses it for the same task. One other aspect of the Striped Possum is noteworthy, namely that when disturbed it gives rise to a very disagreeable smell. Just like the Skunk of America and the Zorilla of Africa, the Striped Possum has both a very striking coat and a very striking odour. The two things seem to go together.

In figure 4 we see another tiny Possum which measures 7 centimetres long, plus another 8 centimetres for its tail. Notice its long, pointed muzzle and its big, black eyes, which are good for seeing at night. Its common name, Honey Possum, tells us at once what it lives on. This little animal, which is so like a mouse, lives in trees and drinks honey from flowers. It has adapted very well to this form of life, so that its teeth are almost non-existent. Its tongue is long and it has a sort of brush, which, with its long muzzle, allows it to reach right into flowers. Thus it eats all the honey, pollen and the little insects which it can find there. In fact *Tarsipes* has adopted the same form of life as have the birds— the honey eaters. Like them it builds a round nest of dried twigs in the branch of a tree. It moves in the trees during the night with an extraordinary agility; its little hands possess opposing thumbs, and its long, prehensile tail acts like a fifth hand. So it scurries about the trees, jumping over little gaps and clinging to the leaves, nibbling at every flower it passes.

THE KOALAS
(Phascolarctos cinereus)

Following page:

1. THE COMMON WOMBAT
(Vombatus ursinus)

2. THE HAIRY-NOSED WOMBAT
(Lasiorhinus latifrons)

One of the best-known animals is the famous little marsupial bear of Australia, the delightful Koala. The exportation of this animal or even of its fur is strictly forbidden. It is therefore impossible to see one of these alive outside Australia, except in one place. The zoo at San Diego, U.S.A., is the only one in the world outside Australia which has living Koalas. The reasons for this prohibition are many. Firstly, it is a measure for protecting Koalas from extinction. Koala fur is extremely beautiful and the animals were starting to disappear, simply because of the demand for their pelts. Secondly, this animal lives exclusively on the leaves of varieties of Eucalypts, which generally do not occur outside Australia—except in the Botanic gardens of San Diego.

While the Great Kangaroos, the Platypus and the Echidna attract the attention of amateur Zoologists throughout the entire world, it is the Koala which most attracts their affection. It is certainly not a bear, except for some external resemblance; it is a marsupial, because it has a pouch which opens towards the front. Its great attraction is that it resembles the toys which children for years have played with, and which they have loved.

The Aborigines have always loved the Koala, and they did not kill it. The word signifies in their language, "that which does not drink". The arrival of Europeans in Australia produced a terrible massacre of the poor Koala, which was discovered in 1802 by the Ensign, Barallier. It was scientifically described when anatomical specimens from expeditions arrived in London at the beginning of the nineteenth century. From then on, it was greatly sought after by people who loved good furs. Hundreds of thousands of Koalas were killed. After the first world war, a law was introduced giving total protection to these animals, which were starting to disappear. Thus the species was saved at the last moment.

Actually, the Australian authorities have worked very hard to conserve the Koalas. They created vast reserves for them, and healthy Koalas were put into areas which they had inhabited in other times and from which they had disappeared. Rangers were appointed to watch them and the suitable Eucalyptus trees were planted where these were lacking. In addition, many intensive studies have been undertaken to improve our knowledge of the life of Koalas. These have resulted in many unusual facts being discovered.

For example: each Koala eats more than 1 kilogram (2 pounds) of Eucalyptus leaves each day and this means that the animals have to change their trees quite frequently, especially when the animals are numerous. Now the Eucalyptus leaf, when it is young, may contain quite large amounts of prussic acid, which is a deadly poison. It has been found that the Koalas know very well just how much of the young leaves they can eat and they supplement these with the less tender, large leaves. Thus they prevent themselves from being poisoned, but this greatly complicates keeping them alive in zoos. Some Australian zoos and some private scientists in Australia have obtained permission to keep these animals in captivity. The Koala adapts very well to these conditions and becomes quite tame.

It is a slow animal, very peaceful and very quiet. It can kill and wound, because it is well-armed with claws, but it does this very rarely. Its beautiful, thick, soft fur never contains parasites, because the odour of Eucalyptus oil, with which it is impregnated, discourages them.

The Koala seems to live for about 20 years. It reproduces every second year, starting when it is 4 years old. The baby is almost always single, except for rare cases of twins. It measures 15 to 18 millimetres long at birth and weighs 3 to 5 grams. The gestation period is 30 to 35 days. Once it is born, the baby Koala instinctively makes the same movements as do all the other marsupial babies; it slowly crawls to the marsupial pouch, which is quite close and then attaches itself to the nipple which is in it. The nipple expands and it is then impossible to remove it from the tiny mouth of the baby without damaging the mouth. For 2 months it stays there without opening its mouth again, but receiving milk regularly for 5 minutes every 2 hours. At 2 months the baby Koala is well formed and starts to climb out of the maternal marsupium. It then measures about 18 centimetres long. Once it leaves the mother, it does not immediately eat the Eucalyptus leaves; instead, the mother gives it material which she vomits up and which is specially made for him. It is a sort of soup made of pre-digested leaves. The young Koala eats this rather strange food for about a month.

It does not finally leave the maternal pouch until it is 7 or 8 months old. Until it is 12 months old, it continues to sleep between the arms of its mother, and is carried by her, either on her back, or clinging to her breast. It is sometimes almost crushed between her body and the tree-trunk which she is climbing, but without its apparently suffering any harm.

On the following page we see two varieties of Wombats —the Common Wombat and the Hairy-nosed Wombat. We know only four species. These animals are becoming more and more rare. They are about 1 metre long and weigh about 40 kilograms (90 pounds). They are nocturnal, peaceful and inoffensive and are easy to tame. They make enormous burrows and eat only vegetation, roots, stalks, leaves, mushrooms, etc. The first variety of Wombat, which is shown on top of page 36, is the most prolific.

THE TASMANIAN DEVIL
(Sarcophilus harrisii)

Australians usually call him "the Tasmanian Devil", or occasionally, "the Badger". He has perhaps a somewhat diabolic appearance, being about 50 centimetres long, and black with some unusually placed white spots. He has a slightly hairy tail and a large, hairy head, stiff whiskers, and claws which are as well developed as those of crocodiles. With all these he has a resolute appearance and a brave nature, together with an irresistible appetite for living or dead flesh. He had disappeared from mainland Australia well before the arrival of Europeans, but was discovered in Tasmania. Very soon the fowls of the Europeans received visits from the Devils, and so the war commenced. The Devil is happy to eat anything he can get and so he was easily taken in traps and with poison-baits. Moreover, because of his extraordinary strength, he defends himself well against dogs and so hunting him gave interesting sport. People dug him up from the deep burrows, which he makes rather like Badgers do. The result one can easily see— the *Sarcophilus* has become rare in spite of having relatively large numbers of babies. Each year the female has up to a maximum of four babies, which remain quite a long time in the maternal pouch. This is quite a large purse of skin, with the round opening placed towards the rear. In this the young ones find two pairs of nipples.

Adding to the bad reputation of the Tasmanian Devil is the fact that, contrary to the other marsupials on the continent, (with the exception of the Tasmanian Wolf), he has a voice which he uses to good effect. He snarls rather like an angry dog and howls in a very characteristic way, which intimidates his foes. He also scratches and bites savagely and has astonishing power in his jaws. If an adult Devil is captured, he can escape from a very solid cage. A few people have brought up a baby and say that they tame quite well and do not then have the bad character which has been attributed to them.

At present it is necessary to go to the wildest regions of Tasmania to have a chance of seeing one of these animals, especially as they leave their burrows only by night. He chases all sorts of diverse prey, such as lizards, snakes, birds, little marsupials of all sorts, sheep when he can find one, and even fish and crustaceans He likes eating anything, and his appetite is as well developed as is his strength. In this regard one can recall another ferocious eater the "Glutton", which is another name for the Wolf found in the north of the American, Asian and European continents. Each of them is possessed by this urge to indulge in a fury of killing and eating, which makes them hated by the inhabitants of the regions where they are found. If they are tamed, both of them prove to be interesting companions. But who would dare to domesticate a Wolf or a Tasmanian Devil apart from some crazy Zoologists?

THE TASMANIAN WOLF, or THYLACINE
(Thylacinus cynocephalus)

The Tasmanian Wolf, or Thylacine, is probably the strongest carnivore of the Australian continent. However, there is a legend or a tradition, which tenaciously affirms that a sort of tiger has been seen in Australia, of which we have no paw-print, no skin, and no authentic photograph. Until such a large cat has been authentically identified, it is the Tasmanian Wolf which must remain as the strongest marsupial carnivore. But at the moment it is one of the rarest animals in the world, and perhaps this species must be considered extinct. An expedition of young French Zoologists explored the south-west of Tasmania in 1967 and found no trace of the animal. On the other hand, it has been reported that recently an observer travelling in a light plane has seen the Thylacine in an almost impenetrable forest which covers this region of Tasmania.

The Tasmanian Wolf need not be confused with any other animal. It is quite large, 1.80 metres long, with a very elongated head, having large teeth in its jaws, which can be opened at a very great angle. Its fur is grey, tending towards a light-beige colour and marked by 16, 17 or 18 bands across the back, extending down to the beginning of a long tail, which thins down to a point. The paws are quite short, but the shape of the rear ones is very characteristic. They are somewhat like those of the Kangaroo. The Tasmanian Wolf is capable of standing on these paws and proceeding by a number of rapid hops; at least that is the reputation which has been given to this animal, because it is a very long time since anyone has observed it running. It is practically impossible to obtain a photograph of the Tasmanian Wolf in its natural state. Those which do exist show the animal in a cage and date from a long time ago.

Robert Dallet has drawn the animal from two different angles, one with its jaws widely opened. It should be added that it has 5 large claws on each paw and that it walks on the pads of its paws. Like all the Dasyurids which have been mentioned so far, the female Tasmanian Wolf has a marsupial pouch in the form of a folded piece of skin, which is open towards the rear and which contains two pairs of breasts. We do not know exactly how many babies she has, but it is thought that she does not have many. This aggravates the effects of the war which for a long time has been waged against the species. Older observers have said that the Tasmanian Wolf has a cry halfway between the meouw of a cat and the bark of a dog. Thus they can be recognised from far away. Local stories have it that the Tasmanian Wolf could climb trees quite well, and that the animal never attacked man, nor did him any direct harm. If so, this would make the animal even more impressive.

However it could easily beat any dog no matter how large, just as the Wolf can do so in northern regions. Nevertheless, it has been said that the introduction of

the wild dog, (the Dingo), into Tasmania has been one of the causes of the almost complete disappearance of the Tasmanian Wolf, because it seems incapable of fighting a pack of Dingos. The Thylacine hunts alone and, because it is not very fast, practises a form of hunting which is well known. This consists of following a faint scent and carrying on until its prey is exhausted, at which stage the Wolf kills it.

The Thylacine eats whatever it can. It is particularly attracted by sheep, which caused in its time the fury of the sheep-breeders, and so the death of the Tasmanian Wolves. It is even capable of overcoming the spines of the Echidnas and, through persistence, of chasing down Wallabies.

We have listed the various circumstances which were the cause of the almost complete disappearance of the last, or perhaps the nearly-last of the Tasmanian

Wolves. We have also mentioned the French Zoologists who tried for several years to find traces of the animal. This south-west centre of Tasmania is densely covered by what is called "scrub". This is a type of vegetable carpet which forms above the ground between the low branches and the high roots of trees. It is impossible to walk across the ground through the scrub, and it is equally difficult to walk on top of the scrub itself for more than three paces. It is in this region that the last Thylacines probably lived, in impenetrable caves, as refugees from more accessible regions. Zoologists looking for the last traces of the Thylacines have perhaps the best hope of success around Mount Lodden and Lake Saint Clair. But they will have to hurry, because if there is one species of animal menaced by extinction, it is the Tasmanian Wolf and, unfortunately, it is irreplaceable.

THE MUSK RAT-KANGAROO
(Hypssiprynodon moschatus)

The Musk Rat-kangaroo belongs to a group of Kangaroo Rats, comprising ten species, of which two are almost extinct. While the Musk Rat-kangaroo of Queensland is the first and the smallest of the long series of Kangaroos, it also forms a sort of junction between these and the Possums.

The little animal is characterised by a naked tail and by the presence in its "hands" of a clawless thumb, which can be opposed to the other digits, as one can see in the drawing at the top right of this page. One can see also that if the thumb of this animal does not have a claw, the second digit has 2, and that there are only 4 digits on its hand. This is a specific characteristic which we will find again soon. Robert Dallet's drawing shows a Musk Rat-kangaroo with two babies putting out their heads from the opening of the maternal pouch. This good mother, sitting down comfortably, is enjoying eating some leaves, which she holds to her mouth with her hands, because her thumbs allow her to grasp things quite well. We will see later other Kangaroo Rats and will frequently find similar attitudes adopted by them, both for carrying their young and for handling their food.

All these little beasts are very similar and it is necessary to look at them very carefully if one is not used to differentiating them. We have taken great care in this volume to show the small differences which will help identify them. This is one advantage of drawings over photographs, and we have lost no opportunity to show such differences.

Altogether the Kangaroo Rats are very restless little animals. They know very well how to use their long hind-feet to jump about in all directions. The males are quite aggressive fighters, but they are all headed for extinction, because they are chased by dogs and foxes which were so unfortunately let free by the first colonists, and which now have swarmed over the whole of Australia.

These animals, the Kangaroo Rats, have also for a long time formed one of the favourite foods of the Aborigines. These primitive people have no equals in finding the faintest trace of a track, a nest or a hole. Because of this, and because the Aborigines had almost nothing else to do but hunt and fish, the numbers of Kangaroo Rats were much reduced under their incessant attacks. The arrival of Europeans with their domestic animals and their excellent weapons rapidly worsened the situation of the Kangaroo Rats, which had already been in only a precarious equilibrium for a long time.

THE RUFOUS RAT-KANGAROO
(Aepyprymnus rufescens)

With the *Aepyprymnus* we have another type of Kangaroo Rat. It does not have the opposable thumb like the Musk Rat-kangaroo, but it uses its hands almost as adroitly. Another difference is that it does not have the naked tail of the Musk Rat-kangaroo, but on the contrary has one covered with the same fur as that which covers its body. The habits of life of the two animals are not greatly different and they are becoming equally rare for the same reasons. Another reason for their disappearance is the habit of the Aborigines, when hunting, to set fire to the scrubby regions where these animals live. Each year many of them are killed but each year the number killed becomes less because the animals themselves are becoming fewer and fewer, until eventually they will disappear.

It is important to understand and be concerned about what is happening. The animals which disappear may perhaps seem without importance, but in the end it is one part of the heritage of all humanity which will have vanished. For example, in a country like France, there are now hunters who kill no more than one Hare per year, but this means that they will kill some fifty in half a century and one is tempted to think that soon there will be no more Hares. Now the Hare forms part of the folklore of France, of its legends and its tales, just as the Kangaroo Rat forms a part of those of Australia. We are beginning to know the details of these animals by talking to the Aborigines who still retain their traditions.

But one should not become concerned only for the Kangaroo Rats and the Hares; all the wild fauna in Australia, as in France and other countries, find themselves involved in this process of destruction by hunting, by bush-fires, by massive doses of pesticides, and by the proliferation of other species which are better adapted or more resistant. Thus we risk leaving to our children waste-lands, empty of many varieties of wild animals, without many bird-songs, and occupied solely by domestic animals. Of this possible future, the best that one can say is that there would be little inspiration or beauty in the spectacle of this ravished Nature.

From the practical or economic point of view the Musk Rat-kangaroo of Queensland and the *Aepyprymnus* add nothing very important to the Australian continent. The poorest sheep offer much more from the point of view of economics. But an Australia which has lost all its native animals would no longer be Australia, and we would regret it if sooner or later most of the animals playing in the natural theatre of Nature were to disappear.

One can understand hostile attitudes towards dangerous animals or those which are pests. However, the little animals shown on these pages are neither dangerous, nor pests, nor do they communicate diseases, nor steal grain, nor kill nor even wound people. To preserve them, we as a community must make some minor sacrifices in our standard of living, e.g. by paying more in taxes or receiving less in benefits from the government for the taxes we already pay. The loss to Australia, as well as to the world would be considerable, if these animals should vanish for ever.

1. THE BURROWING RAT-KANGAROO or BOODIE-RAT
(Bettongia lesueur)

2. THE BRUSH-TAILED RAT-KANGAROO
(Bettongia penicillata)

3. THE EASTERN RAT-KANGAROO
(Bettongia gaimardii)

We continue our study of the Kangaroo Rats with these species, which are brought together under the genus name of *Bettongia*. They are all very similar, with the same small size, the same fairly long fur, the same large head, the same short ears, and the same kangaroo-like attitude, with long rear-feet and small hands.

They have exactly the same habits of life as do the animals in the preceding pages. They are equally as rapid and lively, and they bound about in the same fashion. They are also very quarrelsome, especially when the males fight over the females. There has always been a tendency for Australians to name their animals after European species, to which they bear some minor similarity. Thus these *Bettongias* are sometimes called "rats" or "rabbits". In fact their size does vary between that of the rat and the rabbit, but they are not rodents in the normal sense of the word, but are solely herbivores. They thus occupy the place of the small herbivorous animals in the Australian fauna.

There is one extraordinary development of one of these species (*Bettongia penicillata*), which is called the Brush-tailed Rat-kangaroo. While this animal is not arboreal, it has succeeded in the course of its long evolutionary history in making its tail not only into a prehensile organ, but into a fifth "hand" in the true sense of the word. It is drawn here, (figure 2 on page 43), carrying a bundle of grass, held by this famous tail. The tail, like the rest of the animal, is covered with greyish-brown fur.

This is a fact which continually astonishes people who are interested in animals. We all know very well that the Australian animals are the most ancient in the world and that they live in the most ancient continent. The word "ancient" can also signify "archaic" with a slightly derogatory sense. An "archaic fauna" is a fauna which is to some extent badly adapted, retarded or has not evolved. This is generally agreed. However here is one of these species of animal, firmly embedded in the evolutionary history of the country, which has developed a most useful thing, which otherwise is found only in some of the most evolved animals, the monkeys, with their prehensile tails. Now the prehensile-tailed monkeys usually use this feature for helping them climb and cling to the branches of trees. It is not really a true "fifth hand", but more a supplementary point of suspension. Certainly some of them, e.g. the *Lagotrix*, are capable of handling small objects with their naked tails, but such monkeys are rare.

And here is the ridiculous little Kangaroo Rat, ancient and primitive, which has evolved in the marsupial "blind alley" and yet can do something which all the superior mammals have not been able to do. It has discovered a marvellously rational use for the prehensile tail. With it, it gathers grass and transports it, either for eating if it is fresh, or for making its bed if it is dry. The Brush-tailed Rat-kangaroo is certainly ancient; it is certainly not archaic.

Robert Dallet could not miss such a good chance of showing an animal so beautifully endowed in action. There is also another Kangaroo Rat with the same ability (*Potorous tridactylus*), which is shown on the next page.

The Eastern Rat-kangaroo as with *Potorous platyops*, which is shown on page 45, is now considered definitely extinct and is no longer part of the Australian fauna. It is shown in figure 3 on page 43, but this is a sort of homage rendered to its memory. However it is never possible to say definitely that a species is extinct, because happily from time to time a species which was considered extinct is found one day by accident in some particularly wild region.

1. THE LONG-NOSED RAT-KANGAROO
(Potorous tridactylus)

2. GILBERT'S RAT-KANGAROO
(Potorous gilberti)

3. THE BROAD-FACED POTOROU
(Potorous platyops)

There have been several references to *Potorous* in the preceding pages and it has been said that certain of them have disappeared. In fact, one has to be a very good Naturalist to tell the *Bettongias* and the *Potorous* apart. Both these types of Kangaroo Rat resemble to some extent the mammalian rats and the rabbits, but with a kangaroo-like form. They are herbivores, not rodents. In olden times, when the European Rabbit was so unfortunately introduced into Australia, people noticed that the *Bettongias* and *Potorous* were seen with them and some, e.g. the *Bettongia lesueur*, lived with them down their burrows.

Robert Dallet has shown three species. The third is one of the animals which is now extinct in Australia, but its disappearance is so recent that hope remains that it may be rediscovered one day in some particularly isolated region. All the *Potorous* have a rather long, furry tail and a pointed muzzle. Some have a prehensile tail. They all have particularly short front legs, and when they bound along, one has the impression that these legs do not exist, because they are held so tightly against their chest. It is in this particular position that our artist has drawn Gilbert's Rat-kangaroo at the bottom of page 44.

The colour of these little animals is a rather greyish brown, lighter underneath. They are found everywhere. The Long-nosed Rat-kangaroos are particularly prevalent in the eastern part of New South Wales, between the Richmond and Irrawa Rivers, with a predominance in the rain forests and their edges. Some species of *Potorous* have taken the same path as the *P. platyops* and are rapidly disappearing. One does not know what is the best thing to do to protect them, because it seems that in the end it is simply the presence of people in Australia which is causing their disappearance. The only solution would be to remove the people!

As was the case with the *Bettongias* and as we shall see with the Wallabies, it is necessary to take care when identifying these animals, because they have an astonishing number of different names The long-nosed Rat-kangaroo is called *Potorous tridactylus* in Latin (meaning Three-fingered Potorou) which is its scientific name, but it has been given another name in Latin by other scientists. It is also commonly called "the Potorou" with the addition of various other names, which vary from one part of the country to another. There are even more difficulties when translating these names into the common names in another language because of the different Zoological conditions in different countries. This is frequently found not only with these animals, but with all animals which live in Australia.

The disappearance of the *Bettongias* and the *Potorous* has resulted in a disequilibrium in the balance of nature. The food which they would normally eat is being eaten by other imported animals and they have to look for a new source of food. Thus they are forced to search for other flora to eat, which brings a change in their distribution and this is followed by a change in the distribution of the carnivores, which are obliged to look for other prey following their disappearance. Once more the study of Ecology teaches us that nature is a complicated inter-related structure, and when one part of the chain is lost, the entire structure is disturbed. Unfortunately these ideas have only recently come to the fore, too late to allow us to arrest these processes; return is impossible.

1. BLACK TREE-KANGAROO
(Dendrolagus ursinus)

2. DUSKY TREE-KANGAROO
(Dendrolagus bennettianus)

3. LUMHOLTZ'S TREE-KANGAROO
(Dendrolagus lumholtzi)

4. WHITE-FACED TREE-KANGAROO of NEW GUINEA
(Dendrolagus matschieii)

The Dendrolages are a group of Wallabies which are specially adapted for living in trees. They sometimes descend to the ground, but most of their life is spent in trees. Essentially they eat leaves, little green shoots, berries and fruits more than they eat grass, but they do like ferns and bracken. Robert Dallet has shown four species on these pages.

The Dendrolages are about 60 centimetres long. They all have something of the same body-shape. They have the same rather long muzzles which are somewhat pointed and rather thick. They have the same little ears, and the same variable colouring, which is sometimes bluish and which is a help in camouflaging them among the vegetation. While they have the strong hind-feet of the Wallabies, they have front-paws which are well-developed and have very strong nails which help them climb amongst the trees. In the insert below figure 1, you can see the unusual arrangement of the digits on the paw of the Black Tree-kangaroo; there are 3 digits, but the shortest one has 2 closely set nails. It seems that the animal uses its digits to scratch through its fur, removing anything which is lodged there. They thus become a sort of two-toothed comb.

The Dendrolages, when they come down on the ground, are as good jumpers as the other Wallabies, but when in the trees they show themselves to be marvellous acrobats. They jump from branch to branch and from tree to tree with an incredible ease. They can jump many metres horizontally and allow themselves to fall more than 15 metres in height without appearing to suffer. One can say that the Dendrolages are the Australian animals which take the place of monkeys, which are not found on this continent. There are these differences however—the Dendrolages do not have a prehensile tail as do the monkeys of South America, and they are silent. These factors make them unlike all the monkeys of the world, which are creatures that generally make a great deal of noise. One finds the Dendrolages in the wooded regions of the north-east of Queensland, except that they are starting to become rarer. As usual, the nomenclature, or naming of the animals, is a little bit complicated and the safest names to use are the scientific or Latin names. These are normally quite definite, except of course when the same animal has been independently named by two different scientists. This can cause chaos, until it is finally straightened out.

1. THE BANDED HARE-WALLABY
(Lagostrophus fasciatus)

2. THE EASTERN HARE-WALLABY
(Lagorchestes leporides)

3. THE LITTLE ROCK-WALLABY
(Peradorcas concinna)

The Hare-wallabies are very similar to the Kangaroo-rats but they are a little larger, a little more rapid, and a little finer in their form. They have less-prominent canine teeth and rather more elongated ears. They were extremely abundant when first discovered, but have become rarer and rarer ever since, and are now on the road to an early extinction, as we will discuss later. We show here three varieties of this group of beautiful animals.

The first is easier recognised, because of the bands on its back and on the rear of its tail, while the other Hare-wallabies have a more even-coloured fur. The Banded Hare-wallaby was discovered early in 1699. The second, which also lived on the grassy and bushy plain, is perhaps completely extinct. It is likely that it was unable to compete against the large numbers of sheep which lived in the same regions. It is not a very forceful animal and could not frighten the sheep away from its food, not even when it was starving. In addition the Hare-wallabies are not good at reproducing rapidly.

Each couple has only a single baby and this occupies the mother's attention for a long time. They are sensitive, timid, easily frightened animals, which escape by bounding beautifully and gracefully away at the slightest danger. However, the Hare-wallabies were quite incapable of resisting the inroads made into them in all regions by European hunters with their dogs, and by imported foxes. The competition by sheep and the use of traps and baits of all sorts further diminished their numbers. The Hare-wallabies also had the bad fortune to be a form of game which was found extremely appetising. In the face of all these difficulties, what could they do but disappear?

The third animal shown on these pages is a little more elongated than the Hare-wallabies and is approaching another group which will soon be considered, namely the Rock-wallabies. For this reason Robert Dallet has shown *Peradorcas concinna* perched on a large outcrop, which allows it to look over the surrounding country. With its little front-paws tucked against its chest, the small, reddish animal watches for danger and is ready to spring off with a great, desperate bound, driven by its resiliant, long hind-legs. Like the other little Wallabies whose conditions for survival have become more and more difficult, *P. concinna* has adopted a nocturnal way of life and refuses to let anyone see it in daylight. But this change in its habits of life is not always for the best. The masses of dry rocks where it likes to live are also the homes of numerous snakes, (Pythons, among others),

which greatly appreciate this small, tender game which has arrived on their doorstep. There is a great desire for life among the Australian animals, but the action of man which is exercised in favour of a few as against the others is very rapidly fatal for those who already have trouble in surviving.

The *P. concinna* are called Little Rock-wallabies to differentiate from the other Rock-wallabies which we will discuss shortly. It is necessary to remember that the Australian continent contains about two dozen different species of Wallabies. It is always extremely difficult to recognise one from the others, unless one has an extremely precise description.

People are frequently confused between the Wallabies and the Kangaroos in the numerous zoos where they are shown because, while the Wallabies are generally more attractive than the Kangaroos, the latter's name is much better known. We shall see that there are very few species of animals which can truly be called "Kangaroos". Kangaroos are much larger than Wallabies, although both animals are built on the same lines. The form of these animals with their small front feet and very large rear ones is almost unique to Australia. The only other animals with this general silhouette are some of the giant reptiles, the Iguanodons and the Brontosauruses.

1. THE NORTHERN NAIL-TAILED WALLABY
(Onychogalea ungiufera)

2. THE CRESCENT NAIL-TAILED WALLABY
(Onychogalea lunata)

3. THE BRIDLED NAIL-TAILED WALLABY
(Onychogalea fraenata)

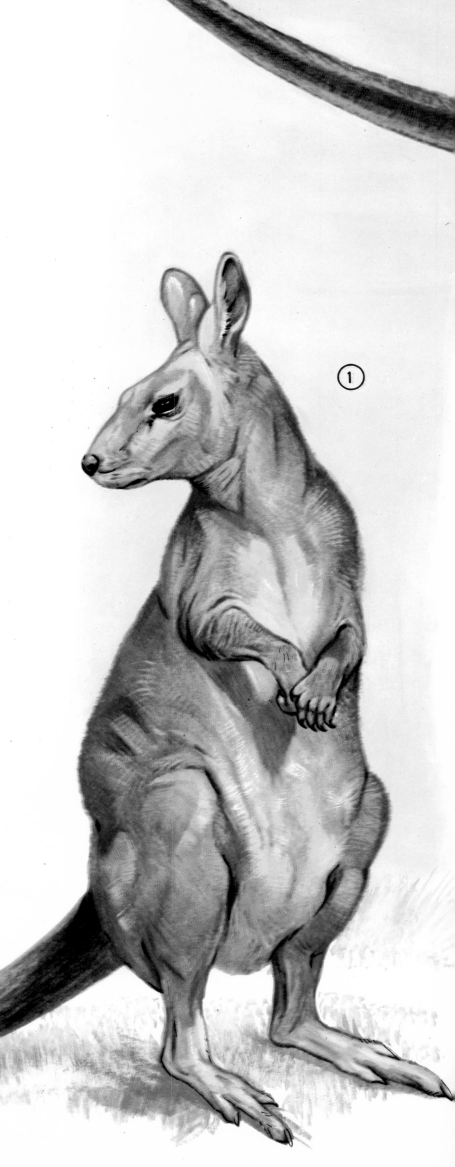

The Onychogales are small to medium-sized Wallabies, of which we know only three species, which resemble the Wallabies of the Plains and the Rock-wallabies. They are most attractive animals, very pleasantly proportioned, with well-developed ears. They bound with a beautiful motion as one can gather from the drawing at the top of the next page.

Regarding the first of these species, we should go back and look at pages 32 and 33 at the bizarre arboreal animal called the *Dactylopsila picata*. At the end of his furred tail, which terminates in a black and then white bunch of hair, there is a spur which has been well shown in the drawing, but for which the use is not known. We remarked on page 32 that Lions also have such a spur in the puff of hair which ends their tail. The Onychogales have the same peculiarity and this is the reason why one of their species is called "ungiufera" (meaning clawed or nailed). This is an allusion to this oddly placed nail in its tail and not to the nails which are on its four paws. However, we are obliged to say that the reason for its existence is completely unknown.

The two other species shown have names which are self-explanatory. One can see the pale crescents which border the front and hind-legs of the Crescent Nail-tailed Wallaby. (This is called "lunata" or "moon-like" in scientific terminology.) One can see also the bridle on the other—a white stripe which goes from its ears to its chest and around its front legs, giving the appearance of a piece of harness.

The beauty of the Wallabies is very similar to that of the Kangaroos. They have well-developed front-legs, armed with curved nails, and beautifully muscled hind-legs, with enormous claws. Their tails serve as a third leg when they are seated, and greatly help to maintain their balance when they are bounding.

Finally, do not forget that the fore-paws, even if they are not true "hands", have opposable thumbs which enable them to gather food and transport it to their mouths. This is a great advantage for a timid animal; it can eat and at the same time maintain a watch for its enemies. A sheep cannot do this.

51

1. THE SHORT-EARED ROCK-WALLABY
(Petrogale brachyotis)

2. THE YELLOW-FOOTED ROCK-WALLABY
(Petrogale xanthopus)

3. THE BLACK-FLANKED ROCK-WALLABY
(Petrogale lateralis)

4. THE BRUSH-TAILED ROCK-WALLABY
(Petrogale penicillata)

Here are four of the small Wallabies. They are agile, can climb and can jump. They live neither on the plains, nor in trees, but in rocky outcrops where they take the place which is taken, in other continents, by some of the monkeys such as the Barbary apes of North Africa or the Baboons and other Cynocephales of East Africa.

These animals are nocturnal. At night they come down from their bare rocks to eat the grass of the plains and to drink water, which they need to do only once every few days. However, it has been noticed that in sancturies, where they are protected and where they have nothing to fear, they revert to the diurnal habits of their ancestors; i.e., they feed and drink during the day.

The Short-eared Rock-wallaby is the most thickset and stocky. The Yellow-footed Rock-wallaby is also called the Striped-tailed Rock-wallaby. One sees these wallabies in a small region of New South Wales, to the north of Broken Hill and in the adjacent region of South Australia, down into the Flinders Ranges and in the Adelaide Hills. They are also handicapped by the severe competition of domestic animals, including imported goats which have become feral, i.e. have taken to the wild.

The Black-flanked Rock-wallaby is drawn here in the process of eating. It is shown with a baby, which is already capable of walking, in its marsupial pouch. The Brush-tailed Rock-wallaby is very pretty, with its dark fur. We see it with a young one drinking from its mother, from the nipple in the ventral pouch. These animals live in the rocks in small groups, principally in New South Wales, where the Jenolan and Warrumbungle National Parks have a considerable number. More solidly built and more imposing than the other Rock-wallabies, this one is easily distinguished by its dark face, which has a white band running from its muzzle towards its ears.

All the Rock-wallabies are inoffensive animals. Graceful and soft, they are never aggressive and are good parents, very happy with life and full of vivacity. They use their hands for gathering and eating food and for caressing their young ones. They sometimes sit day-dreaming on a tripod formed by their back legs and their long tail. One can only be glad of a country which has the good fortune to possess such enchanting animals. One hopes that the survivors will continue to be happy and multiply.

We mentioned above that the imported Foxes are the bitter enemies of the Wallabies and are no doubt responsible for the worrying diminution in their numbers. Even so, the Wallabies have done well in resisting the Dingoes (the wild dogs imported by the Aborigines) and the Pythons (which live in their rocks), as well as their Aboriginal neighbours. Nevertheless the intervention of a new enemy has severely handicapped these poor animals which are incapable of overcoming the wiles of the European Fox. The animals which the Fox catches in Europe have learnt ways of avoiding him, but the animals of Australia have not had the hundreds of years necessary to allow them to evolve such protective reactions. Thus the arrival of European predators has been the final blow to many species.

①

1. THE RED-NECKED PADEMELON
(Thylogale thetis)

2. THE QUOKKA
(Setonix brachyurus)

3. WILCOX'S PADEMELON
(Thylogale wilcoxi)

4. PARMA'S WALLABY
(Wallabia parma)

The *Thylogales* are often named Pademelons, from the Aboriginal word "Pademula", which was used for a whole series of edible Wallabies. These have been much appreciated by the Aborigines. While the Wallabies are usually considered as elegant, graceful animals, the Pademelons are funny clowns. They have a big, fat stomach, a rather serious expression, both astonished and naive, and they certainly would set no records for fast running.

All the *Thylogales* are becoming rarer, without doubt because they have only a single baby at a time. They were discovered very early in 1626, when they were very numerous, living in bushy plains or amongst the trees.

The *Thylogale thetis* is well known in Australia because these animals were very numerous around Sydney and because the early settlers used them for food. They are now quite rare. One variety, which was called the Red-necked Pademelon, was seen between Bundaberg and Moruya.

The Quokka is easily recognised because of its short, stocky silhouette and big tummy, and because of its grey-blue colour. Once it was very numerous in Australia but it is now most abundant in Rottnest Island, near Perth, where it was discovered in 1658 by Volckersen. It is from this island, which has never seen Foxes, that the Quokkas are sent to all the zoos of Australia. Recent studies have shown that when the female Quokka has the misfortune to lose her baby, a second embryo, which was fertilised at the same time as the first, starts to develop as a replacement. This occurs without need of further fertilisation by the male. This peculiarity may explain the resistance of the Quokkas to extinction.

Wilcox's Pademelon lives in the humid forests of New South Wales in the most northern regions. Redder and darker than the Quokka, in its life-style it is midway between that of the Quokka and Parma's Wallaby. This latter is drawn here with one of its young, which is probably too big to fit into the maternal pouch. One can see the round and "cuddly" body of this animal, which seems to be almost smiling, and which has a short, furry tail.

People who know these gentle animals well say that they are quite simple to catch and that they do not need a very great space in which to live quite happily in captivity. This is also true of other animals which closely resemble them, and which we will find on the next two pages.

Efforts to protect the threatened animals of Australia are based on the following principles: first, one studies with care the needs of each particular species, then one looks for a region which has the right climate and the right food, and then the Government declares this to be a sanctuary or a park. As many animals as possible are then captured, care being taken not to hurt them, and released in their new homes. For preference these are on an island or a piece of land which is surrounded by an impassable barrier, if necessary a fence. When the animals have started to multiply and are growing too numerous for the capacity of the reserve, the surplus numbers are caught and sent to zoos in Australia or overseas and to new sanctuaries. The results are often quite spectacular and animals which were nearly extinct are able to multiply and become quite numerous. In this way the Koalas and the Platypuses, for example, have been saved.

OTHER THYLOGALES

1. THE RED-LEGGED PADEMELON
(Thylogale stigmatica)

2. TASMANIAN PADEMELON
(Thylogale billardieri)

3. COXEN'S PADEMELON
(Thylogale stigmatica coxenii)

4. A NEW GUINEA MOUNTAIN WALLABY
(Dorcopsis mulleri)

Here are four more kinds of *Thylogales*. We could add to this list the Darma (or Tammar), a Pademelon which has become rare, *Thylogale eugenii*. This would by no means be enough to complete all the species in this family, using "family" in a loose sense.

We can see that the Red-legged Pademelon has white marks on the sides of its face and around the bottom of its ears. The Tasmanian Pademelon is shown with a baby sticking its head out of the pouch. Coxen's Pademelon is in the process of gathering grass. The *Dorcopsis* has a very short, dark coat. These four types are less frequently caught than the Pademelons shown on pages 54 and 55. They are more like true Wallabies than those we have met so far.

Before the arrival of the first Europeans, with their dogs, foxes, sheep, goats, rabbits and other forms of foreign animals, *Thylogales*, Pademelons and Wallabies lived quite tranquil lives. They had learnt for a long time to beware of Aborigines with their boomerangs, they knew the tricks of the Dingo and avoided the regions where the great Pythons might lurk. Certainly the less agile, the less sensible and less intelligent were killed by their enemies. But equilibrium was maintained between these deaths and the births of new young ones. But once the Europeans arrived, the female *Thylogales*, because they only have one baby at a time, were not able to keep pace with the deaths. This is one of the fiercest laws of nature and will inevitably lead to the disappearance of a species.

There are still other enemies for the little Wallabies— the different varieties of marsupial cats, of which we have talked earlier, the Tasmanian Wolves and Devils. But these decreased in numbers, as did all the other animals, so these could not be held responsible for the disappearances of the various species. Perhaps only the Dingo, which we will consider soon, and which of all these animals is the only one without a marsupial type of reproduction, has always been favoured against the natural Australian fauna. But he arrived in Australia thousands of years ago, so a new equilibrium had been reached which lasted until the Europeans arrived.

As we have said before, but it is helpful to say it again, Australia remained an isolated continent which was separated from the rest of the world. The result was that the animals of the rest of the world evolved along different lines from those which were followed in Australia, where an uninterrupted line of development of marsupials was followed.

④

LOTS OF WALLABIES

**1. THE RED-NECKED
OR BENNETT'S WALLABY**
(Wallabia rufogrisea bennetti)

2. THE GREY-YELLOW EURO
(Macropus isabellinus)

3. THE BLACK-TAILED WALLABY
(Wallabia ualabatus)

On these two pages are shown three varieties of true Wallabies, which are sometimes called Wallabies of the Plains to distinguish them from the preceding species.

In Greek, "macropus" indicates that the animal has large legs. We will see how this description applies very well to the Great Kangaroos. But our Wallabies of the Plains are similarly proportioned. In fact they are exact replicas of their large cousins, but a little smaller. They seem to have been taken from the same mould. We will examine thirteen varieties on the next pages before we look at the two varieties of Great Kangaroos.

The difference between all these Wallabies lies in such details as the thickness of their fur and its colour. Some animals have little interest for the makers of fur coats, either because their colours are unattractive or because their fur is too short; other Wallabies have almost disappeared because women love the beauty of their fur for coats and stoles. While Wallabies of all species were extremely numerous all over the country at one time, they have been massacred with great intensity for two centuries. It seemed that the supply was almost inexhaustible. It is only in modern times that there have been cries of alarm, because we have realized the numbers have fallen so much that in some cases it may be impossible to return to the earlier situation.

The three Wallabies shown here are of medium size. The first, which is called the Red-necked Wallaby, is about 1 metre long. It has been shown by Robert Dallet with a small one in its pouch. The infant is already capable of running by itself, but it loves the comfort of this large, furry pouch in which it has lived for the first few months of its life. The Red-necked Wallaby lives in the wooded hills of regions of New South Wales. One sees him on the road from Newell near Pilliga.

The Grey-yellow Euro is a beautiful colour, while the Black-tailed Wallaby is darker with lighter cheeks and a beautiful black muzzle. It is shown amongst the short grass of the hills where it lives. It has eaten grass for thousands of years, but this is now being used for sheep. This would not be serious if the sheep were not so plentiful, and becoming even more so. Still, Australia has made most of its money from wool, so one can understand their spread. Until very recently, however, it seemed that nothing could oppose the spread of sheep, which threatened to one day occupy the entire country. While we must deplore the present low wool prices, we can be pleased for the many native animals which will be helped by the reduction in the numbers of the sheep.

①

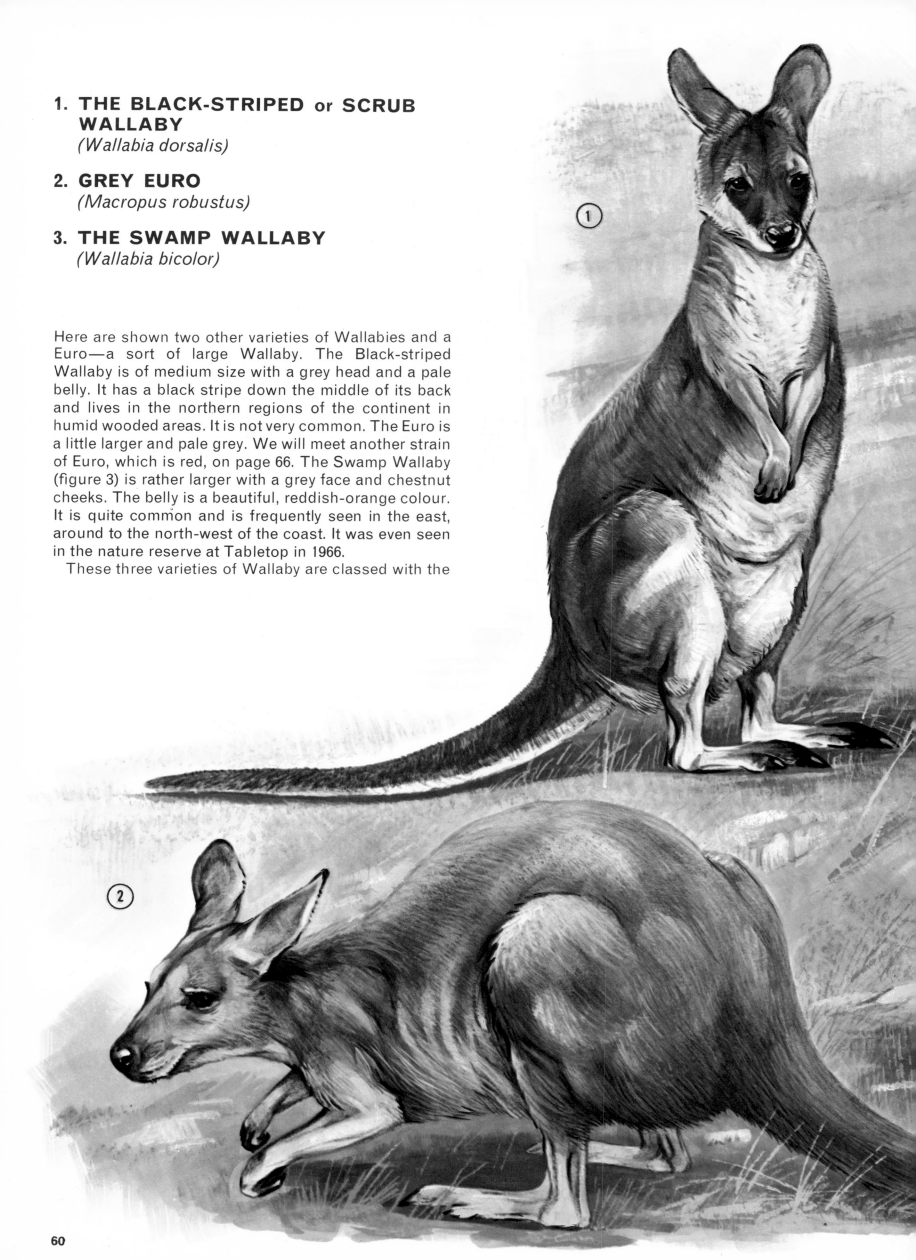

1. THE BLACK-STRIPED or SCRUB WALLABY
(Wallabia dorsalis)

2. GREY EURO
(Macropus robustus)

3. THE SWAMP WALLABY
(Wallabia bicolor)

Here are shown two other varieties of Wallabies and a Euro—a sort of large Wallaby. The Black-striped Wallaby is of medium size with a grey head and a pale belly. It has a black stripe down the middle of its back and lives in the northern regions of the continent in humid wooded areas. It is not very common. The Euro is a little larger and pale grey. We will meet another strain of Euro, which is red, on page 66. The Swamp Wallaby (figure 3) is rather larger with a grey face and chestnut cheeks. The belly is a beautiful, reddish-orange colour. It is quite common and is frequently seen in the east, around to the north-west of the coast. It was even seen in the nature reserve at Tabletop in 1966.

These three varieties of Wallaby are classed with the

others called Wallabies of the Plains. This title is a little vague, but means that they are found throughout most of Australia where the trees are of medium size and the grass grows to various heights. The Wallabies of the Plains are totally herbivorous and they must eat a lot of grass each day to live. To this grass, no doubt, is added a certain amount of green leaves from the bushes that they encounter. Most of these animals live in groups comprising a few individuals or several dozen. These are scattered through the thin bush and seem to ignore each other but nevertheless still maintain contact one with another. Thus if danger becomes apparent, one sees all the Wallabies become alert together. They dash off in the same direction, progressing by huge, supple, silent bounds, resembling the movements of a ball of rubber bouncing on a hard surface.

The Wallabies and the Kangaroos do not run in a zig-zag fashion like the Kangaroo-rats and Jerboas. Knowing how fast they can run, they flee in straight lines, without realising that they thus form an easy target for their pursuers, which run them down. While the Great Kangaroos, which we will discuss shortly, are particularly hated by the sheep-farmers and are hunted for the large quantity of meat which they give, it is for their fur that the Wallabies are killed. And one knows, unfortunately, that in this area supply is always less than demand.

1. BERNARD'S WALLAROO
(Macropus bernardus)

2. THE QUEEN EURO
(Macropus reginae)

3. THE REDDISH EURO
(Macropus erubescens)

On these two pages there are three more kinds of Wallabies. In fact one does not really know if one should call these animals Wallabies or Macropods. Those who prefer the second name say that these animals are midway between the true Wallabies and the Great Kangaroos, the Grey and the Red. Other Zoologists use the term Macropod (from the Greek signifying "large feet" or "large legs") solely for the two very large animals, and call the slightly smaller ones Wallaroos. Thus in books about the fauna of Australia these animals have been described successively as *Macropus bernardus*, *Macropus reginae* and *Macropus erubescens* and are commonly called Wallaroos and not Wallabies. If then Scientists have trouble in naming the animals, it is even more difficult to decide what should be the correct common name.

In fact the Wallaroos attain a good size—up to 80 kilograms for *Macropus robustus* of the Blue Mountains. They can defend themselves very well when attacked, even clawing, wounding and killing dogs. Since they are generally found in rocky regions, pursuing them and hunting them is difficult. They are quite crafty. They allow one to approach them a little more easily than do the timid Wallabies and do not disappear quite as quickly on taking fright. They are becoming rare and the menace of extinction lies over them, as it does over the other large marsupials on the Australian continent.

The *Macropus bernardus* has a dark fur. It is slightly smaller in size and lives in the North, especially in the area of Arnhem Land. The *Macropus reginae* is of medium size, a fawn colour and is common in Queensland. The *Macropus erubescens* is a little smaller. It is found in the centre and in the south of the continent. Its Aboriginal name is "Euro" which many people are in the habit of also giving to *Macropus robustus*, the greatest and darkest of the six varieties of Wallaroos. Other writers do not agree with this, but they confuse a Wallaroo with a Wallaby, and vice-versa. Later on we are going to look at *Macropus robustus*, the Euro, which forms a junction between these animals and the two Great Kangaroos.

Wallaroos in general eat leaves and bark. The rocky regions where they live offer them very little water to drink. They therefore slake their thirst with the pith of trees which they obtain by scratching into them, and by eating the bark of young trees. These animals are less common in zoos because their rather difficult characters make them hard to keep in captivity.

To differentiate the Wallaroos and to avoid confusing them with the other closely related marsupials, it is necessary to take account of their robust bodies, comparing these with the stockier forms of the earlier animals we have discussed. The drawings show the animals crouching on their very strongly developed rear-legs. As well as this, the Wallaroos have a hairless region around their nostrils, which can be seen on the drawings. They generally have a beautiful, short coat. This, however, is considered less beautiful than those of some Wallabies.

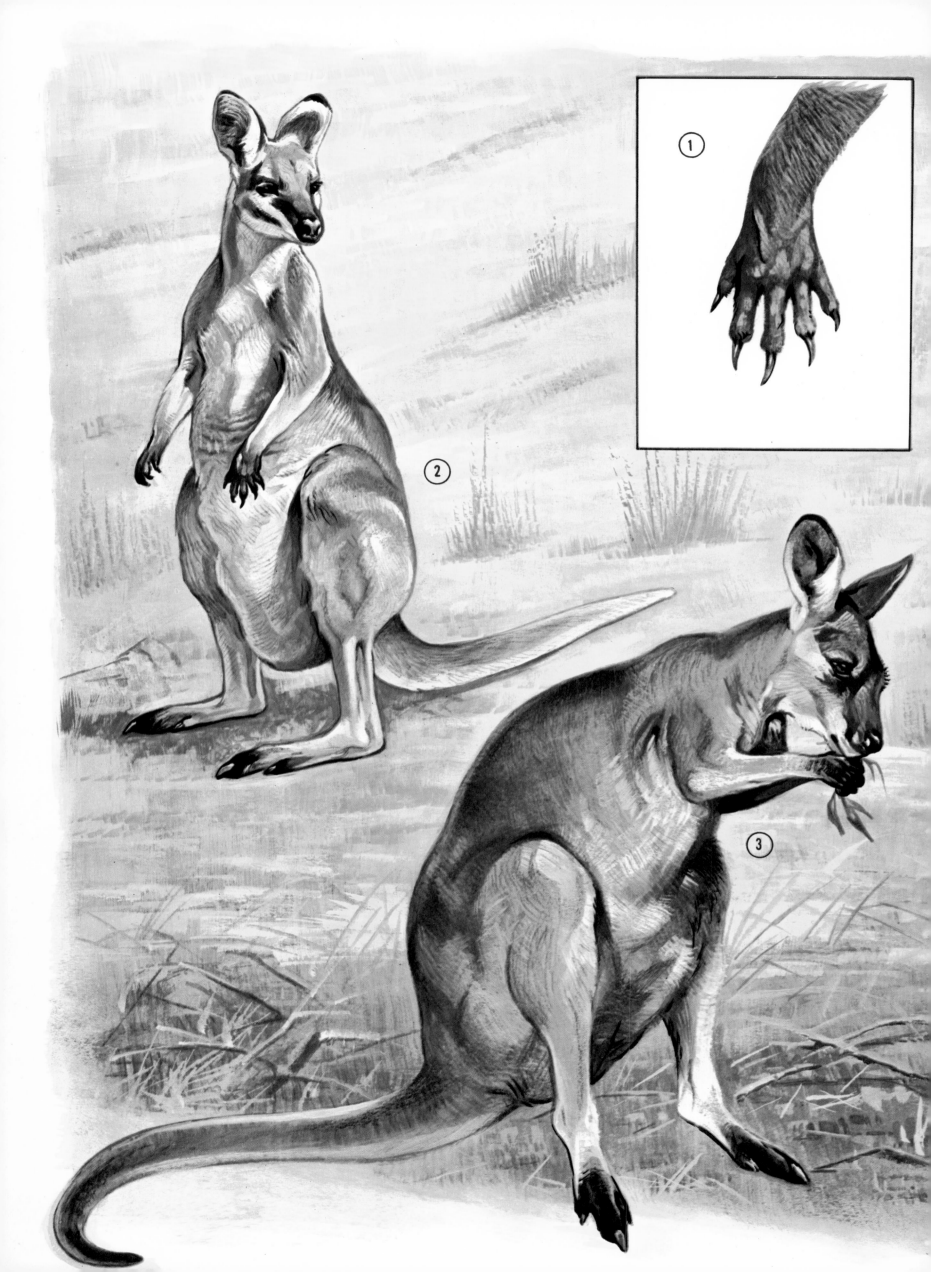

1. A WALLABY'S HAND

2. THE PRETTY-FACED WALLABY
(Wallabia parryi)

3. THE TASMANIAN RED-NECKED WALLABY
(Wallabia rufogrisea)

4. THE AGILE WALLABY
(Wallabia agilis)

Here are three more Wallabies, the last which we will be discussing. The Pretty-faced Wallaby, so poetically named by the early settlers who liked it very much, is a quite large animal. It is grey on its back but paler on its belly. Its ears are brown at the base and its face easy to identify by means of the alternating bands of brown and light-grey. It has beautiful eyes which remind one of gazelles'. It lives in quite large groups and is perhaps seen slightly more often by day than by night. The tail is very long and its hands, which are very supple, have strong, pointed nails. But it is a very gentle animal showing very little aggressiveness. This character trait, allied to the beauty of its fur, has made it rare because it is one of the most hunted of all the Wallabies. It is a miracle that it has been able to resist this pressure and still exist. It is found throughout Queensland and in New South Wales. It loves the woods and the forests. Some people call it the Whiptail Wallaby.

Tha Tasmanian Red-necked Wallaby, which is clearly depicted here, is also quite big and one can see how well it can use its hands. The Agile Wallaby is more finely built, which fact allows it to run and turn rapidly. In the drawing we can see it grooming its front limbs with great licks of its tongue. One can understand why it is said that the Wallabies are very clean animals.

This drawing also shows us that the centre of gravity of the animal is placed over the middle of a rectangle with the base composed of the lower part of the legs where the entire portion of leg touches the ground. Thus this animal can be just as stable balanced on two legs as other animals are when balanced on four.

THE EURO
(Macropus robustus)

The Euro, as we have said in the preceding pages, is characterised by a size intermediate between the other Wallaroos and the Giant Kangaroos. It often weighs more than 80 kilograms. This large animal, which is aggressive and dangerous in the case of old males, has a thick coat and long nails, which are blackish on their upper surface. It lives in the mountains which border the coasts of New South Wales and Queensland, where it is quite common without being abundant.

Another variety, grey this time, is seen in the central regions of the west. This is shown on page 60. In both cases it inhabits dry regions, with many rocks. On these the rough pads on the hind-feet of the Euro do not slip.

For a long time, practically until 1882, Euros were confused with Giant Kangaroos and were considered to be a sub-species, more or less better developed. At this time, the Zoologist, Gould, who specialised in Australian animals, described and named the Wallaroos. (Their name has an Aboriginal origin and comes from "Wallaru", which the Aborigines used for all the Great Kangaroos.) Gould considered they were in the genus *Osphrantus* which other specialists now call *Macropus*.

Like all animals which have large pointed ears, Wallabies, Kangaroos, and Wallaroos use these ears to listen in all directions. Often one ear acts independently of the other, as is shown in the drawing below. It is certainly a very characteristic attitude of this wild and defiant species. One can see again the naked, tender, hairless upper part of the muzzle which is a characteristic of the Wallaroos. Finally, the drawing by Robert Dallet shows us certain features, which are almost imperceptible (a partly closed eye perhaps), that indicate that the Wallaroo has a disagreeable temperament compared with the Wallaby. Its expression has, in effect, certain warning signs which make one decide not to stroke it.

THE GREY KANGAROO
(Macropus giganteus)

The Grey Kangaroo is the largest of them all. It is over 3 metres long (even 3.30 metres) from its muzzle to the extremity of its long tail. It is frequently called the "Great Grey".

The Grey Kangaroo lives in scrubby, bushy regions. It loves the typical Australian plain where there are many Eucalyptus trees. It eats basically grass and leaves. There are a number of sub-species: one has a black head, one a darker colour; one is found only in Tasmania. Always the animal has the same character—a little crafty, and confident in its speed (which is between 40 and 50 kilometres per hour). This can save it from hunting-dogs and horses, but not the rifle-bullet. But if the Great Grey has been extensively hunted for its fur, for its meat and for sport in the last century, it seems now that popular opinion is favouring it. Many people are taking care to ensure that it survives. In fact, it is perhaps the least menaced of the Great Kangaroos.

States in which it is protected most are Queensland, New South Wales and Victoria.

However, in spite of all, it is on the point of extinction. This is probably less because people are killing it, than because the pastures where it grazes have been taken over by sheep. The Great Grey Kangaroo is very choosy about its diet and does not like to live on poorly-producing country, which is all that is left for it.

Sooner or later it will be necessary for men to sacrifice some large fertile regions to make national parks where the rare species of Australia can be preserved. These living monuments are as precious and as difficult to remake as any which have been constructed by humanity in the course of the ages.

The Grey Kangaroo is rarely seen in captivity outside Australia. Grey Kangaroos used to be used as "boxers", because they could use their front feet to fight, like a human boxer, and gave much amusement to people at fairs and shows. But they are rarely seen now either in captivity or in the wild. They have been recorded on film and one can see these magnificent animals, 3 metres long bounding at 50 kilometres an hour with great leaps of up to 12 metres, when the country is favourable. This is a spectacle which no other animal can offer.

THE RED KANGAROO
(Megaleia rufa)

We will give two pages to the giant Red Kangaroo, because it is the best known and most characteristic of all Australian animals. The male is commonly known as the Red Kangaroo, while the female is called "the Blue Flyer". The old males, which are very good boxers, are capable of rapidly killing two or three dogs, or a man armed with a stick. The old man roo must be approached with care, if he is not friendly.

The size of the Red Kangaroo is slightly less than that of the Grey. The young, of which there is usually only one, are born after a gestation of 35 to 40 days. They then continue growing until their death, which can be up to an age of 15 years. The mother Kangaroo makes a path by licking her fur with great sweeps of her tongue, when the baby is born, so that it finds its way to the pouch. On page 65 we show a new-born baby Kangaroo in its natural size. Figure 1 shows one at the time it enters the marsupial pouch, where it will remain for months. It takes a nipple between its jaws and does not let it go. In figure 2 the baby is shown as it is at the age of 2 months, again natural size, which is about 7 centimetres long. In figure 3 we see it after 50 days. We will not go through the details of this life which the young kangaroo passes for such a long time in the mother's pouch. It is similar to that of all the other marsupials and like them, the young Red has this peaceful refuge of a comfortable pouch. When it has grown a little and is capable of surviving for some time outside, it browses in the grass alongside its mother. The sight of a baby kangaroo as it emerges from the pouch, passing first its ears, then its head, and then its front legs over this natural balcony, is always extremely amusing and touching.

On page 68 is shown a large old male Red Kangaroo in a fighting posture. He is balanced on his enormous tail, with his front feet held up ready to box, ready to claw, ready to disembowel, and ready to give a furious double-blow with his rear legs. In fact, ready for anything.

THE BIRTH OF A KANGAROO

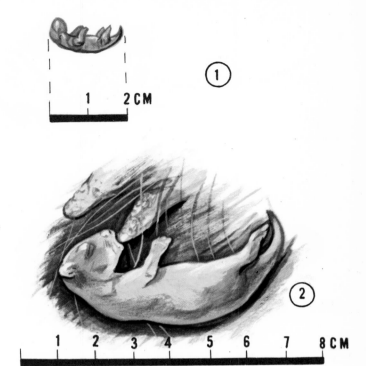

Because the Red Kangaroo is such a great fighter when cornered, he has caused the death of many people, and is accustomed himself to fight to the death. He has also drowned many dogs which have chased him, by going into the deep water and then holding their heads under until they drown. The hands which he uses to box with can be used also to claw and to grab; his nails are extremely powerful and can disembowel a dog or a man. One must not forget that behind his relatively lightly muscled arms, he has a mass of more than 100 kilograms ready to fight, to hit and to kill.

However, these facts have not been enough to discourage hunters. The great Red Kangaroo, who eats grass, is in competition with the sheep-farmers, who can readily kill him with rifles. This species has been menaced by sheep-farmers, and Zoologists have been obliged to battle on his behalf. Eventually, they have obtained for him promises of national parks, unfortunately encircled by fences. The Red Kangaroo is the enemy of Australian Farmers, just as the Dingo is and just as, in Tasmania, the Wolf and the Devil are.

Two-and-a-half years after the second World War, 2,500,000 Kangaroo and Wallaby skins were exported from Australia. The meat of the Kangaroo is used for dog and cat-food. The demand (particularly in the United States) has increased by 1,200 per-cent over the last few years and the Kangaroos are hunted ten times as much as they should be if the species is to survive. There is a risk that the history of the Red Kangaroo will terminate here, unless adequate measures of conservation can be rapidly established.

In addition, the animal's tail, which is used to make soup and the meat of which is eaten, is greatly desired in numerous countries and has always been liked by Aborigines. It is possible to buy it in nearly all large cities in the world, frozen or in tins. The skin is used in many different industries where its strength is appreciated.

The Red Kangaroo lived in the great pastured plains. The troops of sheep pushed him into the forests, which were impenetrable to them, and into the scrubby regions, where they had more difficulty in passing. In some regions, kangaroo-proof fences keep them out of the better pastures, which are used by the graziers.

The female Red Kangaroo is a very fast animal. Like her Grey cousins, she can bound over considerable distances at 40 and even 50 kilometres per hour. She is also extremely graceful when stretched out on the ground, with her little one playing between her front feet, under her beautiful eyes. This spectacle can be seen in the reserves where these animals can live without being hunted. Outside these reserves, this sight is unfortunately becoming rarer and rarer.

1. A BABY KANGAROO at birth
(actual size)

2. A BABY KANGAROO
about 8 weeks old
in its mother's pouch attached to a nipple

3. A BABY KANGAROO
at about 50 days
(enlarged)

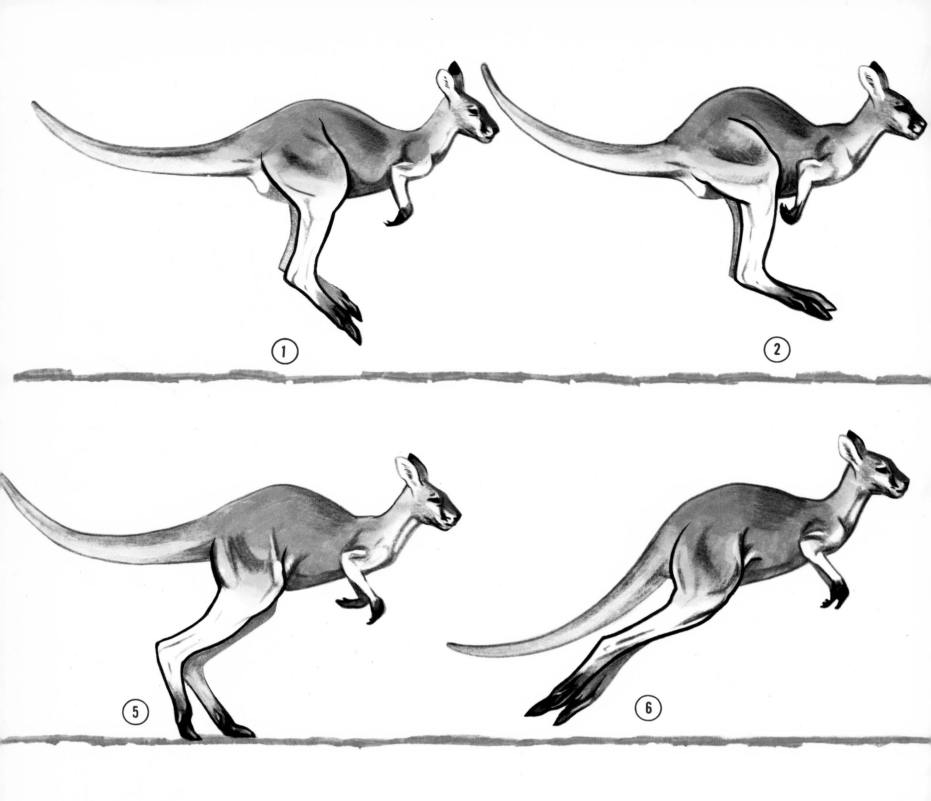

THE GAIT OF THE KANGAROO

The gait of the Kangaroos, especially the Great Reds and Great Greys, is very special. It is quite different from that of all other quadrupeds. Briefly, it is a gait where the back legs move together, while the front ones are held against the chest and assist only slightly by helping to maintain equilibrium. However, the rear feet are not completely in the same line. One is often several centimetres behind the other.

The tail has an extremely important role, because it is most effective in balancing the animal. Without this, the Kangaroo could not bound. When the legs are well forward the tail is well backward, to balance the animal. When the legs are to the rear, the tail comes down so the centre of gravity of the body is centred more over where the legs touch the ground. Thus the tail goes in

the reverse direction to the hind-legs. The 8 drawings at the top of these pages show one complete bound. The larger drawing at the bottom on the right shows all of these superimposed, so that one can see the various positions of the different limbs of the animal. It is interesting that when the rear legs come forward, the front legs go backward, and this also helps to keep the centre of gravity over the point where the feet touch the ground.

In this fashion the Kangaroo starts moving, with little bounds of 2 to 3 metres. He accelerates so that when he is bounding at 40 kilometres per hour, each bound is about 8 metres long. When he obtains his maximum speed of about 50 kilometres per hour, the bounds are about 12 metres long. He can keep this up for a long time, if he is being chased and is fit.

DINGOES
(Canis familiaris dingo)

Chasing an
AGILE WALLABY
(Wallabia agilis)

The Agile Wallaby, which we have talked about already, is a very fast, light animal. He has no trouble in keeping ahead of a dog for thousands of metres. However, dogs can run him down if they keep on chasing him for a long time, because eventually, while they are slightly slower, they can exhaust him. These tactics are used by the Wolves of Europe, Asia, and America, by the Dholes of India and the Hyenas of Africa. The Dingoes chase in a pack, and pursue the same animal until they have run it down. If they can, they will stalk their prey, or ambush it, but since most animals keep a good look-out, the Dingoes normally have to run their prey down.

Unlike all the other larger mammalian animals in Australia, the Dingo is not a marsupial. He is, in fact, the same species as the domestic dog. Thus dingoes have a number of babies, after a gestation period of 60 days. They do not place them in a marsupial pouch, as they do not have one.

This is easily explained; the Dingo is not truly a native animal of Australia. He was introduced in Prehistoric times by the Aboriginal people who came from Asia towards the Australian continent, after passing over various islands en route. The Dingoes were their domestic dogs and were often used by the Aborigines for hunting. This alliance of man and dog has been seen all over the world. The Dingo is descended from the dogs which accompanied the men who came southwards towards Australia. The Dingo is somewhat unlike other dogs and is more remote in character. He will form an alliance with man but is not completely domesticated.

Physically, he is a short-haired, yellow dog with sharp pointed ears. chestnut eyes and quite a short tail. Those young ones which have been taken into captivity have domesticated well and become very agreeable pets, but it is said that this learned attitude is unlike the innate

72

savagery which one finds in the young of these animals from generation to generation. Thus some say that each generation must be tamed afresh, which is quite possible. In zoos he is a slightly timid animal, rather anxious, who continues to walk sadly round and round his cage. In Australia he is an enemy.

Before the arrival of Europeans, the Dingo had already practically eliminated the Tasmanian Wolf and seriously reduced the numbers of many defenceless animals, such as the Numbat. The arrival of sheep, cattle, and fowls from Europe was obviously a happy thing for the Dingo, who killed large numbers of them. For this reason there began a pitiless war with rifles, traps, and poison between the farmers and the Dingoes, although they do not actually attack men. This war continues today, but the Dingo is extremely cunning, defiant, and resistant to attack. Those which remain are hardy and courageous, with a capacity to survive. They always begin to multiply as soon as circumstances move in their favour.

The Aborigines call the Dingo "Warrigal", and he is sometimes so called by Europeans. The origin of the word "Dingo" has not been explained. It has been suggested that it is derived from a slang French word used for a fool, but this is not very suitable for the Australian wild dog which is very far from foolish. The Government pays bonuses for Dingo scalps, but it will be a long time before the last of them are killed or even confined to regions where they can do no harm.

The Dingo has become part of the folklore of Australia in the same way as the Wolf is part of the folklore of Europe. The Wolf has now disappeared from Europe; one wonders if the Dingo will ever disappear from Australia. At present the policy is to confine him, with the aid of Dingo-proof fences, to regions where there are few sheep. If he can be successfully kept where he can do no harm, it would seem to be a pity for this very beautiful dog to become extinct. He may yet be able to be completely domesticated and his nature changed by selective breeding. Since breeding produced the St Bernard, the Poodle, the Kelpie, and the Pekinese, all from the same basic animal, it should not be too hard to retain the beauty and intelligence of the Dingo, while removing his savagery.

①

THE TRUE RODENTS

1. THE HOPPING MOUSE
(Notomys cervinus)

2. THE WATER-RAT
(Hydromys chrysogaster)

3. THE GIANT WHITE-TAILED RAT
(Uromys caudimaculatus)

4. THE FILE-TAILED RAT
(Melomys cervinipes)

In spite of what we have said before, that there are only marsupials in Australia, in fact there are a few small animals which are true mammals and which colonised the country before the Europeans arrived. These are the Rodents, of which there are eighty-five species, to which one should add four other species introduced by man (the Brown Rat, the Black Rat, the House Mouse, and the Rabbit). In fact these four species are characteristically undesirable and have unfortunately multiplied greatly.

The eighty-five other species have been in Australia a long time, in spite of its distance from other countries. They have proved very successful immigrants for they have been assimilated into the life of the Australian continent, and have found roles which they can play in the equilibrium of Nature, without upsetting it and without depriving other animals of their food. The Rodents are an extremely ancient class. They have evolved following the evolutionary directions needed by the roles they play both in Australia and overseas, so that we again have evidence of the convergence of widely separated species, provided their way of life is similar.

On these two pages we see four species of these animals, which are extremely little-known, badly studied, and often confused with marsupials by many observers. The first of these animals has followed an evolutionary path which has given it the classic form of the Hopping Mouse, very like the Jerboas. It has long rear-legs, very small front ones, and a long, furry tail. It is a small animal, dark on the back and light on the belly.

The evolution of *Hydromys* is quite different. It has become a water-rat, but a water-rat of very large size, often being 75 centimetres long. It eats fish, small mammals, and poultry when available. It has short ears, well-developed jaws, and slightly-webbed rear-feet. There are strong nails on its front feet. The tail is very long and its body is thickly furred.

The other two animals have gained, through the ages, features which are not so typical of Desert Rodents. They are of medium size, with few special characteristics, except that they have many young, unlike the marsupials. They are frequently confused with marsupials however.

It is unfortunate that these animals have not been studied very much, for they could make an important contribution to Zoology and the theories of Evolution. Once more this convergence in a form of unrelated animals can be seen. It can also be seen between the Shark and the Porpoise and between the Rat-kangaroos and the European Gerboas. It is due to the fact that two classes of unrelated animals happen to occupy the same ecological niches; that is, they have the same way of life. Of course, these ecological niches must be in different regions of the world, or else the slightly better-adapted of the the two animals would eventually eat the food which the other one needed, so that it would be forced either to change or become extinct.

THE SEA LEOPARD
(Hydrurga leptonyx)

The Sea Leopard is a Seal, a great spotted Seal which is up to 3.60 metres long in the female and 3.20 metres long in the male. It is not as easy-going as most of the other Seals, which defend themselves only if they are attacked. On the contary, it is extremely aggressive and attacks everything which moves, on the land, or in the water. It has extremely well-developed jaws with very long, pointed, canine teeth, which are followed by molar teeth having three very sharp points. The Sea Leopard eats Pelicans, because it lives in the waters and on the islands in Antarctica, and it also eats the young of other species of Seals. It also eats innumerable fish, young Whales and all sorts of birds, which it catches when they rest on the rocks and ice-flows.

One should not be astonished to see this animal added to the already long list of Australian animals. This is because the Australian continent is quite close to the primary region of distribution of the Sea Leopard, which is often seen on outcrops from Macquarie Island and in the southern regions of Tasmania. It is even seen on the coasts of the Australian mainland, notably to the north and south of Sydney and in Victoria.

While it is often seen close to the coast, the animal can also make very long voyages through the sea. It can find there all it needs for survival and, like all sea-going mammals, it can sleep on the surface or even under the surface of the water. (It surfaces automatically whenever it needs to have another breath of air.) Again, like all the other sea-going mammals, it automatically closes its nostrils under the water so that it does not drown.

In brief, the role of the Sea Leopard in the Antarctic Ocean is the same as that of the Polar Bear at the other end of the world. It maintains the natural equilibrium by killing the weak, the injured and the old of other animals and fish and birds. In this it succeeds very well, but for man to encounter it on the ice or in the water is extremely dangerous. It can slide on snow or icy ground with very great speed and is extremely agile in the water, like all the other Seals. It will even take bites out of the gunnels of dinghies if it finds one. It is feared wherever it exists.

Happily it is not very well known these days and few men have seen one. Most people have seen only the gentle Seals, with their beautiful eyes like those of a friendly dog. These are quite unlike the Sea Leopard, which will attack without provocation and considers everything living as a possible prey.

1. THE GREY-HEADED FLYING-FOX
(Pteropus poliocephalus)

2. THE SPECTACLED FLYING-FOX
(Pteropus conspicillatus)

In Australia, there are about fifty species of Bats, which, like those in the rest of the world, are placental mammals and not marsupials. Some eat insects, others fruits, and one (*Macroderma gigas*), which is very large, is a carnivore and eats other bats. Among the fruit-eating species, which are called Flying-foxes, we have shown two which are the most characteristic. These are varieties of *Pteropus,* which is also found in the rest of the world, viz. in South America, Africa, the Indies and Pacific Islands.

The public, which loves to dramatise things, has taken a long time to get out of the habit of considering that these pacific animals, who live only on juicy fruit, are not Vampires. The Vampire is a Bat which is smaller and which has few characteristics in common with *Pteropus*.

Pteropus loves both wild and cultivated fruit, which he prefers ripe and juicy. At evening one can see hundreds, thousands and tens of thousands of these animals spread their great wings (which may measure a metre across), flap their great sails of skin, and fly towards the best orchards and gardens in their neighbourhood. Then, hanging by one or both legs, with their heads down, the bats gorge on bananas, oranges, pawpaws, mangoes, figs, peaches, pears or any succulent, tasty fruit. They may bite into the fruit and use their prickly tongue to scrape at its interior while they suck the juice, which flows down. Then the animals, once more satisfied, fly away and perch again, one against the other, in the branches of the tree where the colony lives. For a while disputes break out; one squeals, one pretends to kill another with his claws, and then all is quiet as they sleep.

The Aborigines sometimes light a big fire under a tree where there are lots of Bats, and hit the asphyxiated animals on the head with sticks or boomerangs. They love the meat, which has a strong, musky odour. It is perhaps disagreeable at first to Europeans. We are probably missing out in this and perhaps if we ate more Bats we could both enjoy a new taste and help the poor fruit-growers of Australia, who live in bat-infested areas, and who regard them as an enemy.

You can see in the drawing the pretty head and pleasant colours of the two *Pteropus*. These are shown in typical attitudes of life, living in a tree. These animals live very well in captivity and become quite tame. If they reproduce, although this is unusual in captivity, the sight of the baby clinging on to the mother's fur, upside down, is very strange and attractive.

SOME GIANT BIRDS

1. THE GENYORNIS
(Reconstruction)

2. THE EMU
(Dromaius novaehollandiae)

3. THE EMU'S EGG

4. THE CASSOWARY
(Casuarius casuarius)

Among the three giant birds, the first, (*Genyornis*), no longer exists. Robert Dallet has drawn a reconstruction from its fossil remains. The true Moa lived only in New Zealand. A similar but unrelated animal the Australian Moa, or *Genyornis,* had a somewhat similar appearance. They have both been extinct for several thousand years. Thus they existed at the same time as the first men, who carved pictures of them in rocks. The *Genyornis* was between 2 and 3 metres in height.

When Australia was discovered by Europeans, there were two species of Emu. The smaller, *Dromaius diemenianus*, existed only on Kangaroo Island in South Australia, but it has become extinct.

The other Emu is met in many regions of Australia. It

is a large bird without apparent wings, but it can run very rapidly. It can give tremendous blows with its strongly clawed feet, and should only be approached with care, even when in captivity. Like the ostrich, it can give heavy blows with its large, hard beak. It is 1.50 metres high to the base of its neck and has an imposing profile. It lives in small groups in dry regions, where it eats all sorts of food, particularly grains and fruits.

The Cassowary is a little smaller than the Emu. One should perhaps say Cassowaries, since there are a number of varieties—at least a dozen. These are not confined to Australia, but are also found in New Guinea and various islands. They are black-coloured birds which have long hair-like fur rather than feathers, very strongly clawed feet, and, on the head, a sort of hard helmet of varying shapes. The head and neck are naked. The neck has a scaly skin, reaching from the head to the breast-bone and of a colour varying between red and blue. In place of wings, the Cassowary has long, rigid and solid spines. These seem to help in opening a path through the shrubs, which it passes through while looking for berries, fruits or grains, and also for small animals, reptiles or mammals, which it spears with its hard pointed beak. Thus the Cassowary is rather better armed than the Emu. Even in captivity, it can be extremely dangerous and attack its keepers. The Aborigines captured them for food and kept them in very narrow cages inside which the poor animals could not move.

Altogether the Cassowary is tending to disappear and certain varieties are considered to be totally extinct. The protective measures which have been taken to look after the giant birds of Australia came too late for these, and there is no hope of bringing back a species which has disappeared. There is chance that some pairs may exist in remote regions, but this is not likely. However, at least the Emu is alive in large, though diminishing, numbers. The birds which we will talk about in the following pages, were also in danger of becoming extinct, but it seems that precautions will be both sufficient and in time to prevent this.

4

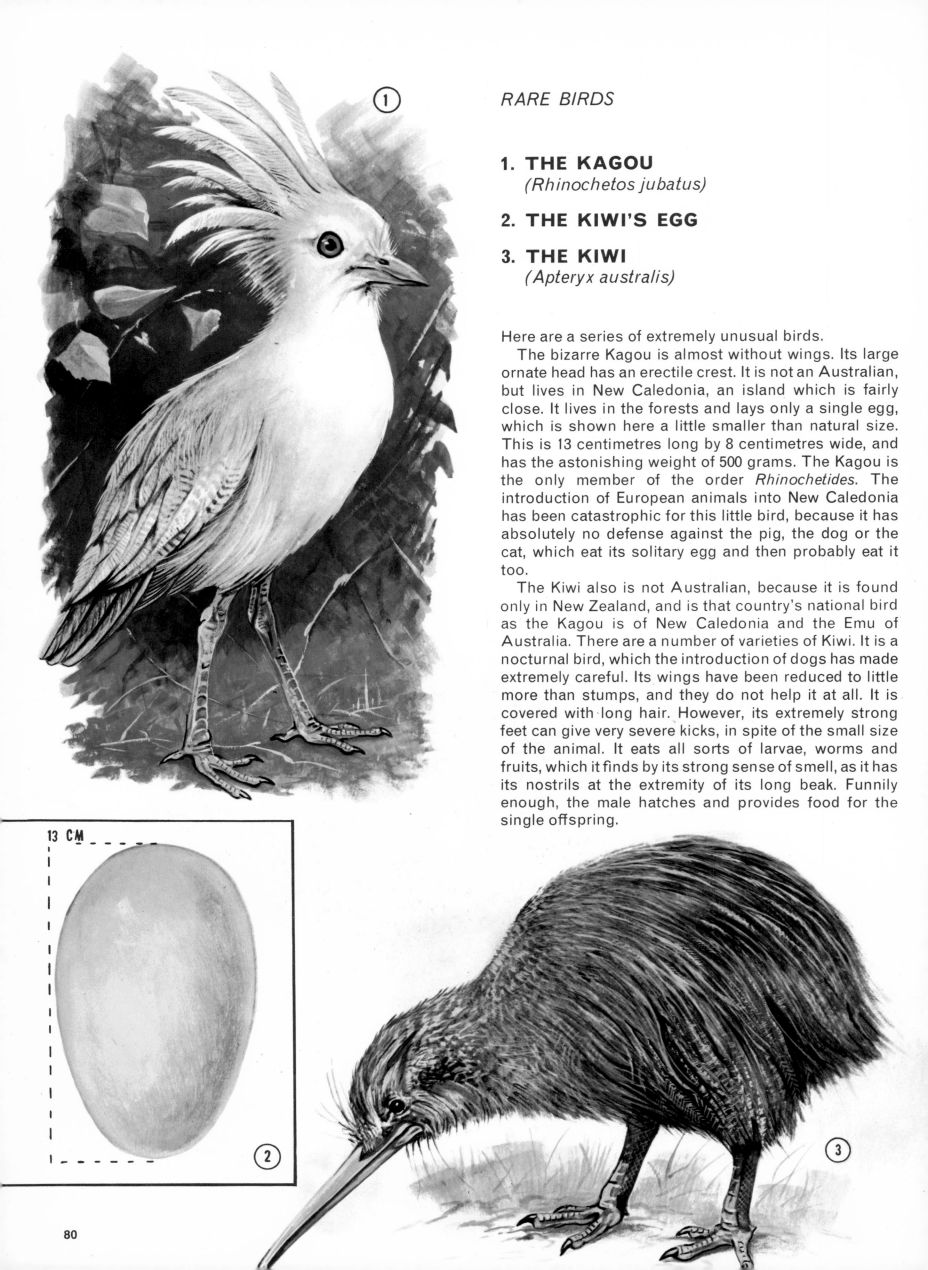

RARE BIRDS

1. THE KAGOU
(Rhinochetos jubatus)

2. THE KIWI'S EGG

3. THE KIWI
(Apteryx australis)

Here are a series of extremely unusual birds.

The bizarre Kagou is almost without wings. Its large ornate head has an erectile crest. It is not an Australian, but lives in New Caledonia, an island which is fairly close. It lives in the forests and lays only a single egg, which is shown here a little smaller than natural size. This is 13 centimetres long by 8 centimetres wide, and has the astonishing weight of 500 grams. The Kagou is the only member of the order *Rhinochetides*. The introduction of European animals into New Caledonia has been catastrophic for this little bird, because it has absolutely no defense against the pig, the dog or the cat, which eat its solitary egg and then probably eat it too.

The Kiwi also is not Australian, because it is found only in New Zealand, and is that country's national bird as the Kagou is of New Caledonia and the Emu of Australia. There are a number of varieties of Kiwi. It is a nocturnal bird, which the introduction of dogs has made extremely careful. Its wings have been reduced to little more than stumps, and they do not help it at all. It is covered with long hair. However, its extremely strong feet can give very severe kicks, in spite of the small size of the animal. It eats all sorts of larvae, worms and fruits, which it finds by its strong sense of smell, as it has its nostrils at the extremity of its long beak. Funnily enough, the male hatches and provides food for the single offspring.

13 CM

(4)

4. THE CAPE BARREN GOOSE
(Cereopsis novaehollandiae)

5. THE BLACK SWAN
AND ITS CYGNETS
(Cygnus atratus)

The Cape Barren Goose was discovered in 1792 by the French expedition led by Entrecasteaux. It is a large, beautiful bird, and quite rare. It has a short beak, and grey and black feathers. It lives in Australia and New Zealand.

The Black Swan of Australia is the counterpart of the Swans of Europe, which represent for many people a symbol of purity, of absolute whiteness. It can be confused with no other bird. It lives on the Australian continent, particularly in the south, and is the emblem of Western Australia, where it was seen for the first time.

It prefers salt water to fresh water, and is a beautiful flyer. Its 4 or 5 large bluish eggs are hatched by the male by night and the female by day, for 35 days. When they are born, the cygnets are greyish, as one can see in the drawing. Then they speedily become black. These are birds with a beauty, nobility and grace which are incredible.

The public is very familiar with the Black Swan because it is present in most of the great zoos of the entire world. The contrast between these birds and the various varieties of White Swans is so extraordinary that a director of any great zoo would feel ashamed if he could not show them to visitors. Fortunately, the Swans are very easily tamed and thus their conservation is simple. Naturally, like all birds living in parks, it is necessary to clip their wings to make sure they do not fly away. This operation does not hurt them, but it is better to have it done by a Vet. It is, however, preferable to clip only a single wing; if both are clipped the animal may be sufficiently balanced to be still able to fly.

(5)

1. THE YOUNG MALLEE FOWL or LOWAN, leaving its sandy nest.

2. THE MALLEE FOWL, covering its eggs.
(Leipoa ocellata)

The Lowans or Mallee Fowls (which are sometimes called Brush Turkeys) live in several different parts of Australia, as well as New Guinea and the Pacific Islands. These large birds are famous among Naturalists because frequently a number of females lay eggs in the same hole, dug by one male. When they have finished, and there may be hundreds of eggs in the hole, the females completely forget about them. The male then starts a busy time because he must cover the eggs with branches and sand until there is a hill sometimes as high as 2 metres. This allows the eggs to develop at their optimum temperature. If the temperature rises during the day, however, the male opens the mound and lets the eggs get some air; when the temperature falls, he covers them again. He continues this without ceasing day and night so that the best temperature is maintained for the eggs. During this time, the poor male Mallee Fowl almost completely ceases to eat because he cannot leave his important job. We do not know exactly how he can be sure when to cover or uncover the eggs, or how he knows their exact temperature, but one thing is certain—he does it extremely well.

When the chicks leave their eggs, they attack their covering, which is of varying thickness, and they separate it and scratch their way to the surface. We can see one doing this in the illustration at the top of this page. It is leaving its sandy nest and is about to start on its new life. This life is very easy for the females, which have no responsibilities, and is full of hard work for the males.

The nests of the Mallee Fowl often receive the attention of bird-nesters, both four and two-legged varieties. Certain reptiles, the Tree Monitors and the Goannas also help themselves. The Aborigines, and then the

3. THE LYREBIRD
(Menura novaehollandiae)

4. THE FRIGATE BIRD
(Fregata ariel)

Europeans, have done the same. When the eggs of the Mallee Fowl are completely fresh, there are enough in each nest to make enormous omelettes. This causes despair in the poor father, who is shown in figure 2 covering the nest.

The Lyrebird is well named, as can be seen from the picture of the male on the right. The long feathers of its tail come down in the form of an inverted antique lyre, but without its strings. It exists throughout the south-eastern coastal regions, in the forests between Brisbane and Melbourne. There are two varieties, and Zoologically they seem far removed from other species. For a long time people hunted the male Lyrebirds for their feathers. Then it was seen that the birds were disappearing because the female lays only a single egg a year, and so they are now strictly protected. This has given very good results.

Not only is the Lyrebird extremely beautiful, but he sings gloriously. He is capable of imitating all the bird songs which he hears and also any noises which are made, both naturally and by men. His repertoire is enormous. Sometimes he seems to sing just to amuse himself. At other times he gives a presentation of his talents of singing and of dancing in an extraordinary fashion, to attract the female. This occurs in a well isolated wood, on about a square metre of land. Very few observers have seen this strange spectacle, but those who have will never forget it. Curiously, the female has very little colour or interest in her plumage.

The large Frigate Bird is one of the best gliders amongst the sea birds. During its mating season the male can be distinguished by an enormous red, inflatable pouch, which is heart-shaped, and is situated under his beak.

④

(4)

BIRDS OF PREY

1. THE SPOTTED HARRIER
(Circus assimilis)

2. THE BLACK-SHOULDERED KITE
(Elanus notatus)

3. THE BROWN HAWK
(Falco berigora)

4. THE FORK-TAILED KITE
(Milvus migrans)

The Brown Hawk resembles the Peregrin Falcon of Europe, because of its size and its way of life. It is a high-flying hunter, who attacks all sorts of little birds, but which helps maintain the natural equilibrium of animal life. The Fork-tailed Kite, which migrated to Australia, is like the Merlins in the rest of the world, with a forked tail like a Swallow, and the useful habit of eating carrion.

The same problems occur to the Australian people as occur to the inhabitants of other continents; namely, should one or should one not protect birds of prey? The same answer is true in all countries. Birds of prey are essentially useful birds and they should be protected with great vigour. If the birds of prey, both nocturnal and diurnal, are to survive, they must be protected from hunters. Unfortunately there are always some foolish people who want to shoot anything which flies or runs, simply for the joy of killing. Others find great joy in watching birds of prey flying with outstretched wings, which hardly move as they plane through the air, ascending and descending, until their keen eyes see some little rodent or rabbit. Then they dive at incredible speed and seize it. On the subject of rabbits, it has been shown that because birds of prey kill so many of them, it is good for the farmers not to shoot these birds, in spite of the fact that some of the larger ones kill some baby lambs. The many rabbits they destroy allow more sheep to graze on a piece of country and more than make up for the few lambs which they take, which are usually the sick and weak ones anyway.

The eye of the bird of prey is extraordinarily acute. It enables it to see small animals, while it is flying very high. It is said that the Brown Hawk can see a mouse moving through grass, even when it is many hundreds of metres high.

In the next four pages are some birds of prey which hunt by day on the Australian continent. On these two pages are shown some birds of prey with short tails, which are comparable to their cousins which live in most places of the world. These cousins are the Buzzards, Kites and Hawks in Asia, Africa and America. Since, of course, these birds fly, these different species have been able to migrate to Australia without difficulty. The Frigate Bird, which we have seen on the preceding page, is capable of flying enormous distances, because its wings are very narrow and very long (more than 2 metres). It flies almost without flapping them, using the air currents. The Frigate Bird has to do this because it does not have the oil, which impregnates the feathers of other marine birds, and so allows them to dive and to rest on the surface of the water without sinking. The Frigate Bird cannot touch the water and must rest on land. To eat, it has to attack other birds and steal the fish which they have caught by diving. For this reason, we know that it is capable of crossing oceans 2,000 or 3,000 kilometres wide without resting. While it can sometimes rest on isolated islands, it normally sleeps whilst flying very high up above the waves.

We will see later on that there are some typically Australian birds. These are incapable of crossing the oceans, so that they could not have come here and colonized the continent, but must have developed in Australia. But to come back to the birds of prey, they, like the Frigate Bird, can have come long distances to colonize Australia.

The Australian Buzzard (or Spotted Harrier) resembles the Buzzards of Europe and has the same habits of life. It likes large stretches of water and fishes in them. The Black-shouldered Kite is quite small in size, about the same as the European Wood Pigeon. It is a marvellous protector of harvests, because it destroys enormous quantities of noxious insects of all sorts, especially the grasshoppers, when they have their great migrations.

1. THE WHITE-BREASTED SEA EAGLE
(*Haliaeetus leucogaster*)

2. THE LITTLE EAGLE
(*Hieraaetus morphnoides*)

3. THE WEDGE-TAILED EAGLE
(*Aquila audax*)

These are the large birds of prey. The Fish Eagle is common throughout all Tropical Asia and Australia. It is an Osprey, as one can tell by its white belly. This is the group of birds to which the Golden Eagle (the emblem of the United States) belongs. There are some in Europe, which are nearly extinct, and some also in other countries, where there are some varieties with white tails. These are usually carrion-eaters. They are certainly capable of fishing, but usually catch sick and wounded fish which have to surface, and which they seize whilst flying. Mostly, however, they eat dead animals washed up on the beaches or elsewhere. Thus over much of the world they have the role of Vultures, and get rid of dead bodies, so preventing possible epidemics. They are useful garbage collectors. In Australia they are particularly useful, since Vultures do not exist here.

The Little Eagle is the size of a large Buzzard and is a great killer of rabbits. As with the other medium-to-large birds of prey, they play their part in helping to rid Australia of this pest.

The Wedge-tailed Eagle has a darkish colour and it plays the same role as do the large Eagles of Europe. He flies at low altitudes, watching for his prey. Look at the drawing, which shows the long claw of his thumb, which crosses those of his other digits. When a Rabbit or a Rat-kangaroo is foolish enough to allow himself to be surprised by a Wedge-tailed Eagle, the Eagle dives towards him and, without slowing down, pierces him with the two long thumb-nails, which are veritable daggers. One can see also the large wing-feathers, which are very supple and can be widely separated. These act as air-brakes and allow him to slow his flight when he wishes, by allowing the air to filter between them.

All the birds of prey are extremely beautiful and very useful. There was no need to be concerned that they might eat-out their prey, and become extinct through lack of food, because they have coexisted for a long time with the animals they eat and have come into equilibrium with them. That was before the arrival of Europeans. Since then the small native animals have become far fewer and the hungry birds take some of the domestic animals—lambs, fowls and even turkeys. However, as mentioned before, it is likely that the eagles kill so many rabbits that they are a help to man rather than an enemy.

THE STRANGE WORLD OF
THE BIRDS OF PARADISE

1. THE BLACK SICKLE-BILLED BIRD OF PARADISE
(Epimachus fastosus)

2. THE LESSER BIRD OF PARADISE
(Paradisea minor)

3. THE GREATER BIRD OF PARADISE
(Paradisea apoda)

In all of New Guinea, and in a small part of the north-east of Australia, there are birds which are quite extraordinary. These are a large number of different species of birds, which are grouped under the general name of Birds of Paradise. On these two pages and on the following six pages, we show some of their many forms.

In general these birds have been hunted for a long time, so that their feathers could ornament the hats of the beautiful women of Europe and America, and thus they have been greatly reduced in numbers. There are an incredible number of Birds of Paradise with shapes and colours which are all extraordinary, but which vary from one to the other. Almost always it is the males who are the most ornate and most beautiful. Their size is usually about that of a pigeon, but their enormous tails make them appear much larger.

On these two pages we can see the Black Sickle-billed Bird of Paradise displaying its plumage in front of its female, then the Lesser Bird of Paradise and then the Greater. The name of the last species contains the word "apoda", and this signifies "without feet". This is a scientific souvenir of the time when people thought that Birds of Paradise did not have feet, because they were bought from the native hunters minus heads and feet. Scientific names are permanent once they have been given; thus they continue to be used, even if they were wrongly chosen.

The first bird described, called the Sickle-billed Bird of Paradise, is not so named because it is a relative of an Ibis with that name, which lives in northern Africa and southern Europe, and in other regions of the world. This Bird of Paradise is so called only because it has a long curving beak like the Sickle-billed Ibis itself. The resemblance ends there, but it was exploited by the first Ornithologist who described the bird. He was European and was reminded, when he saw this bird, of the Ibis which he knew so well.

1. THE BLUE BIRD OF PARADISE
(Paradisea rudolphi)

2. THE TWELVE-WIRED BIRD OF PARADISE
(Seleucides melanoleuca)

3. THE RED BIRD OF PARADISE
(Paradisea rubra)

Following the Birds of Paradise shown on pages 88 and 89, which are some of the best known of the group (but whose ways of life are mostly unknown), here are three more. The first is displaying its feathers in part of its mating ritual, as it hangs by its feet from a branch. The second shows the richness of its beautiful plumage and its strange bib-like attachment underneath its neck. The third shows the complicated form of the long feathers of its tail. These birds nest very high up in the trees and are very suspicious. They do not sing but give raucous cries, which one can hear from far away. These are in strong contrast with the beauty of their plumage.

Of the forty-three species which are known (and without doubt others will be discovered), thirty-eight live in New Guinea or on the little islands between it and Australia. Thus they do not belong to continental Australia, but do belong to the Southern world.

It is as impossible to describe these beautiful birds as it is to describe the marvellous Humming Birds of the American continent. One must look at the drawings by Robert Dallet to get some idea of their form and colours. They are not very big birds, measuring usually between 15 and 35 centimetres. In fact they are cousins of the Crows and, apart from their decorative plumage, are very similar to these and to the Magpies. The magnificent colours are almost always reserved for the males alone; the females are much more soberly clothed. The males display their glorious plumage to attract the females during mating rituals. Some Birds of Paradise show these by hanging from a branch with their head low and the feathers falling over them like a glorious waterfall. One sees an example of this on page 90. It is extraordinary that many species of these birds live side by side in the same regions of the forests, with no mis-mating of the species. Yet the territories of individual birds of the different species overlap.

1. THE ARFAK SIX-WIRED PAROTIA
(Parotia sefilata)

2. THE KING BIRD OF PARADISE
(Cincinnurus regius)

3. THE MAGNIFICENT BIRD OF PARADISE
(Diphyllodes magnificus)

4. THE WAIGEU BIRD OF PARADISE
(Diphyllodes republica)

Here are four more kinds of Birds of Paradise, which have been known for some time in Europe because of specimens taken there. There, their feathers were often used to make small, pitiable little collections, which were kept in boxes made of camphor in the bottom of wardrobes. These were used from time to time to deck a hat or other aricles. We must hope that this fashion never returns.

There is no need to inquire why the Six-wired Bird of Paradise has its name; it is sufficient to look at it to understand. The three others have remarkable shapes which are reminiscent of those of some other birds in South-East Asia and near the Philippines. But it is the size and form of the feathers of the tails and necks which are incredible.

Birds of Paradise do not lay many eggs. There are generally one or two, and the incubation period is 12 to 18 days. But it must be said that we know very little about the way of life of Birds of Paradise. There is room for much fascinating and rewarding study by those who care to push their way through the New Guinea forests.

Birds of Paradise, as far as we know, live mainly on fruits, berries and grains. But it is probable that they also eat insects, worms and larvae of all sorts. The shape of their beak, which is long and strong, makes one think they belong to the group of omnivores, for which everything is food.

In some species the tail feathers are up to 1.5 metres long, which is most remarkable for a bird which is no bigger than a pigeon. One wonders what is the use of these ornaments. It is probable that they are a way of fixing the attention of the females during sexual displays, so that they may recognise the males of their own species and avoid those of other species. Thus the different species of birds, which live together, mixing one amongst the others, are able to keep separate. So species which could originally mate with each other were able to separate and become further and further apart, genetically and evolutionarily speaking. This was also necessary because the times when mating was possible differed, as did the period of incubation, and thus the males and females of different species were not able to mate profitably. Such matings would result in no fertile eggs and hence the disappearance of the various species. Nature is a "racist" who does not approve of mixed marriages.

1. THE SULPHUR BIRD OF PARADISE
(Lophorma superba)

2. THE KING-OF-SAXONY BIRD OF PARADISE
(Pteridophora alberti hallstromi)

3. THE BROWN SICKLE-BILLED BIRD OF PARADISE
(Epimachus mayeri bloodi)

4. THE MAGNIFICENT BIRD OF PARADISE
(Craspedophera magnifica)

On these the last two pages devoted to Birds of Paradise, we find four of the most spectacular varieties. If people had not seen these birds living in their native jungle one would have thought that they had been invented by a facetious Taxidermist.

The name *Lophorma superba* means a very beautiful young lady, and indeed the effect, as can be seen in figure 1, is similar to some of the more extravagant and remarkable dresses which one can see on such young ladies. The bib-like crest of feathers under her neck, combined with the larger crest over her back, give her a most amazing appearance.

Some Naturalists, while being overawed by the forms of these birds, have sometimes given the name of some king or emperor to a species, either from amusement or perhaps for flattery. One can thus find Kaiser Ludovic William amongst their names.

These birds unfortunately cannot live their lives without trouble. For a little time they have been relatively at peace, but they are always menaced by civilization's extending into their forests and by their being hunted by Europeans or by the local people, who also love their beautiful feathers.

The *Craspedophera magnifica* is commonly called the Magnificent Bird of Paradise, as is *Diphyllodes magnificus* on page 92. This confusion serves to remind us that common names may mislead one and it is safest to keep to scientific names, although, as we have seen with the Macropods, even these may be in chaos.

For people who do not live in New Guinea, it is frustrating to think about these birds. They are so beautiful and yet so inaccessible. It seems that they should perhaps live in a park where people could admire them and photograph them. Unfortunately, these birds, together with the Lyrebirds and many others which exist throughout the world, are seen only by explorers capable of going to the places where they live. They do not exist in aviaries around the world.

We should say, in fact, emphasize, that these birds are drawn in their natural colours. Fourteen varieties of Birds of Paradise are not often illustrated in the same volume and the drawings shown here represent a serious piece of work by our illustrator.

1. THE YELLOW-TAILED BLACK COCKATOO
(Calytorhynchus funereus)

2. THE KEA
(Nestor notabilis)

3. THE KAKA
(Nestor meridionalis)

4. THE KAKAPO
(Strigops habroptilus)

It has often been said that Australia and the islands around it are great countries for parrots. A very large number of species of such birds live here.

The first of the birds shown here is typically Australian, although it is not as common as other cockatoos. The Black Cockatoo loves to live in the tops of trees, from which it gives a sad, plaintive cry. It often uses its beak to move around the branches as if it was a third hand, although it is already very adroit with its two hands, which each have two digits opposing the other two. It is a bird which is little known in the rest of the world. Like the White Cockatoo, the Black can attain a respectable age of 100 years.

The Kea, Kaka and Kakapo are parrots of New Zealand. They have become quite rapacious carnivores. They alight on the backs of sheep and tear into them with their long curved beaks to eat the skin and fat of these animals. They also often eat dead animals and serve to get rid of carrion. They are about 45 centimetres long, and live in a nest made of twigs put together on the ground or in a crevice in a rock. They lay 2 to 4 eggs, which hatch in about 3 weeks.

The two *Nestors* live at a quite high altitude in the tops of the mountain ranges. They eat worms and all sorts of insects which they find under the ground.

The Kakapo resembles a little Owl, and like them is nocturnal. This is another good example of the Convergence of Forms. It has been greatly reduced in numbers by the introduction of predatory animals from Europe and is now found only in isolated regions in the South Island. It spends almost all its life on the ground, looking for little animals and plants, which it cuts or scrapes off with its long beak which is formed like a curved dagger. It forms a separate family, of which it is the only species, and is thus very interesting to Zoologists. New Zealand and the surrounding islands form a paradise of unusual animals.

97

1. **THE MAJOR MITCHELL COCKATOO**
   (Cacatua leadbeateri)

2. **THE GALAH**
   (Cacatua roseicapilla)

3. **THE MUSK LORIKEET**
   (Glossopsitta concinna)

4. **THE CRIMSON ROSELLA**
   (Platycercus elegans)

If there is one sort of Australian Parakeet which is remarkable, it is certainly the Cockatoo, of which there are many varieties. We do not show here the best known one (namely the Sulphur-crested Cockatoo, which looks like an enormous camellia) but the Major Mitchell Cockatoo, which is also beautiful with its cap of feathers looking like an enormous Indian chief's head-dress. In captivity the Cockatoos are very gentle birds which know their masters very well and get much pleasure in being caressed. They have good gifts of mimicry, as

have some other types of birds, and can repeat human words very well. Is this intelligence—or just mimicry? If it is intelligence, are they intelligent because they are able to speak, or can they speak because they are intelligent? We must leave this question because we do not know the answer.

One thing is certain—the Parakeets and Budgerigars are the most gifted talkers of all, while the Australian Parrots seem inferior to those of South America and Africa. The Cockatoos in particular are extremely beautiful and rather funny, like clowns, but they do not speak very well and have rather piercing cries. They fly well, but are not very visible because they keep in the trees. Here they live, except when they come down to the ground to look for food and then climb back from branch to branch.

The Lorikeets, which should not be confused with arboreal mammals of Asia called "Loris", are birds characterised by a tongue ending in a group of hairs. With these they remove the juice from fruit, which is their food. These Lorikeets, which originated in Australia, have colonized a number of islands and archipelagos of Oceania—Fiji, Samoa and the Society Islands.

We are going to see a great number of Parrots, starting on page 99. These are all ravishingly beautiful birds, and we shall start with the beautiful Crimson Rosella, shown here. One of the advantages of Parrots in a cage is that they do very well in open-air aviaries, provided they have a warm corner and are sheltered from bad winds when the temperature becomes too low. They reproduce very well under these conditions.

1. THE EASTERN ROSELLA
(Playtycercus eximius)

2. THE NORTHERN ROSELLA
(Platycercus venustus)

3. THE ALEXANDRA PARROT
(Polytelis alexandrae)

4. THE MULGA PARROT
(Psephotus varius)

5. THE RAINBOW LORIKEET
(Trichoglossus haematodus)

A further fourteen species of Parrot are depicted for us on the following six pages. We shall see that all the Parrots have the same rounded head with a high forehead, the same largely-opened, lively eyes, the same curved beak and the same feet, having two digits opposing two others. The wings are beautifully coloured and the tail is usually long, very mobile and acts to balance the animal when it is clinging to a branch. It also acts as the rudder and elevators of an aeroplane do to allow the bird to turn and dive very quickly during its flight. Often many of these birds fly in a group, at which time their aerobatics are astonishing. There is one other detail to note; it is possible to separate the two halves of the beak quite widely, which other birds cannot do. As we have said, the beak of parrots constitutes a form of third hand, letting them move along the trees.

Parrots love to live in large groups, so when caged they should not be isolated. Indeed they get on very well with each other and different species do not quarrel. Parrots of the same species particularly love to live together.

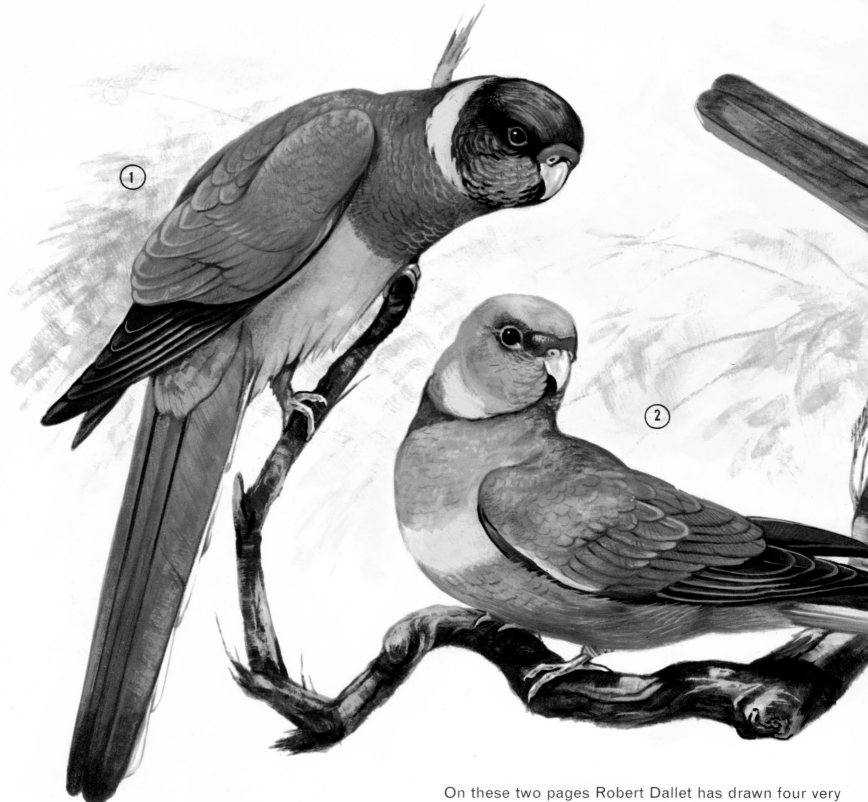

1. THE PORT LINCOLN PARROT
(Platycercus zonarius)

2. THE RING-NECKED PARROT
(Platycercus barnardi)

3. THE WESTERN ROSELLA
(Platycercus icterotis)

4. THE GOLDEN-WINGED PARROT
(Psephotus chrysopterygius)

On these two pages Robert Dallet has drawn four very beautiful Parrots. There is no need to emphasise the extreme beauty of these little birds. The colours of the drawings speak for themselves and there is little need for a supplementary description. All the same, we can briefly describe some of the life-style of these Parrots, which are so typically Australian.

They were first known in Europe about 1840 when the English painter, J. E. Gould, made the first coloured drawings of them. At that time people knew that they flew with rapid changes of course, that they lived generally on grains, and that they lived in groups numbering from less than a hundred to many thousands. It was known that they laid between 5 and 8 eggs once or twice a year, in nests under the limbs of trees, and that the eggs took 18 to 20 days to hatch, and that the young stayed in the nest 6 weeks.

Parrots are migratory, following the ripening of the grains and fruits on which they live. They are capable of learning to speak, as are all the *Psittacides,* but they do not speak as well as the marvellous talkers of South America.

Bird-breeders are very interested in Parrots and have succeeded in breeding many varied strains, which are quite different from the original ones. They have produced many pleasing colours, which give an even higher value to the birds. However, this book deals with wild animals and not with domesticated species. The tame Parrots, which have been reproducing in cages for many generations, have become domestic birds.

One of the things which is most noticeable when one looks at the life of Parrots is their "married love". The most remarkable from this point of view are the famous Love Birds, which are to all effects inseparable. Even if these birds cannot speak as humans do, they speak to each other in a way which resembles chattering, with giggling and little noises which are most affecting for those who hear them. When one comes on a group of Parrots in the wild, one immediately thinks of a crowd of little human beings exchanging impressions and gossip. It is very impressive to hear the diversity of expressions of Parrots compared with those of other chattering birds, who repeat again and again the same notes in the same rhythm.

1. **THE TURQUOISE PARROT**
   (Neophema pulchella)

2. **THE RED-RUMPED PARROT**
   (Psephotus haematonotus)

3. **THE RED-WINGED PARROT**
   (Aprosmictus erythropterus)

4. **THE COCKATIEL**
   (Nymphicus hollandicus)

5. **THE KING PARROT**
   (Aprosmictus scapularis)

We will end our look at these ravishing Australian parrots with five more examples. There are many more species, but those we have shown give some idea of these delightful creatures, which one can meet in their hundreds, as they sit on trees or on the tall shrubs of Australia. Often the parrots are victims of heat-waves and drought. They are not capable of very long flights, and so if water disappears they will speedily die. The different species are maintaining themselves quite well, although some of them have become extinct since their discovery in the middle of the last century. But many remain to charm the bird-lovers of Australia. However, if they are to survive in large numbers, it will be necessary to leave a large area uncultivated where they can find their normal food, which is indispensable to them. Unfortunately, there is competition between birds and men for the use of most land that would be suitable for this purpose.

1. THE RAINBOW BIRD
(Merops ornatus)

2. THE CHESTNUT-BREASTED CUCKOO
(Cuculus castaneiventris)

3. THE DOLLAR BIRD or BROAD-BILLED ROLLER
(Eurystomus orientalis)

These three birds are close relatives of other species which are found in Africa and Europe.

The Rainbow Bird is one of the greatest killers of Hymenoptera (bees, wasps, etc.) that we know. He is very helpful and eats, without flinching, one wasp after another. He is a great aerial acrobat, as he catches all of his food on the wing. His long slender bill is very helpful for this purpose. Because he eats so many bees, apiarists dislike him and often shoot him. In Europe many people are concerned about the disappearance of such birds, as this is followed by an increasing abundance of noxious insects.

The Australian Cuckoo, or Chestnut-breasted Cuckoo, loves caterpillars, like his relatives in other parts of the world. One can say then that he is one of the most useful birds and helps rid this increasingly hungry and famine-stricken world of moths and butterflies in their larval stage. In Australia, as in Europe, it is necessary to protect the Cuckoos. It is not always easy to explain to some shooters (we will not dignify them with the name of hunters) that some birds should be respected.

The Roller of Australia is different from that of Europe, but it is also a great eater of pests. It is an ally which should be appreciated by man. It likes to perch high up on a bare branch, from which it takes off in pursuit of passing insects. Its name is derived from the spectacular ability it has of rolling over and even somersaulting in mid-air during flight.

The presence on the Australian continent of these types of bird, which are so like those on other continents, justifies again what we have said earlier. Birds are quite capable of passing from island to island, and from continent to continent, by flying. This is no doubt what happened with these three birds, which are not specifically Australian. They have been transformed in their new environment, following the Laws of Evolution, in common with all species which are now living or have disappeared.

We will show on the following pages other birds, but not nearly all the birds of Australia, of which there are many varieties. There are, however, fewer varieties than in Asia, Africa or America. These continents have an advantage over Australia by offering a more varied climate and more wooded areas. Birds in the end are dependent on trees, rather than on bushes.

The fifteen other Australian birds which we shall look at in the following pages give us some idea of the bird population of this continent. These birds exhibit many surprising features, as will become apparent when you turn these pages.

1. THE KOOKABURRA
(Dacelo gigas)

2. THE RED-BACKED KINGFISHER
(Halcyon pyrrhopygia)

3. THE SACRED KINGFISHER
(Halcyon sanctus)

4. THE WHITE-TAILED KINGFISHER
(Tanysiptera sylvia)

The Kingfishers of Australia, if they are fishers, are very similar to those in the rivers of other countries. But the most extravagant and extraordinary Kingfisher of all is the Giant Kookaburra. The Kingfishers fly along the surface of the water to capture fish in their long, strong, pointed beaks. They can be seen in figures 1 to 4. In figure 4 can be seen the strange plumage of *Tanysiptera sylvia*.

But to come back to the Kookaburra. It is quite a big bird with an enormous head and a beak which looks like a little dagger. It is also given all sorts of other names, of which Laughing Jackass is perhaps the best known. "Kookaburra" is derived from the Aboriginal name for this bird. While it has the same form as all the Kingfishers of the world, it is the giant of the family. There are two varieties; the one which we see here lives in the east and south of Australia. There is another, with blue wings, which lives in the north and west. This bird is distinguished not only by its size but by other peculiarities. It enters houses and steals brilliant little objects which attract it. It is carnivorous and hunts snakes, even the most venomous ones, rodents and small marsupials. It catches quite large lizards and fears nothing. Finally, this strange bird breaks into peals of laughter on all occasions. It is found all over Australia and can be heard both day and night. It lives well in captivity and laughs just as loudly in zoos, much to the consternation of unwary visitors. Curiously, although he is a Kingfisher, the Kookaburra drowns if he gets too many of his feathers wet by trying to catch prey near pools of water. It seems to have lost the water-proofing oil which the other Kingfishers possess.

1. THE PHEASANT COUCAL
(Centropus phasianus)

2. THE WHITE-BACKED MAGPIE
(Gymnorhina hypoleuca)

3. THE CRESTED GOURA
(Goura coronata)

The Coucal is a sort of Cuckoo with an unusual, long tail which reminds one of a pheasant.

The White-backed Magpie is a beautiful bird, which is black-and-white, and which has the same habits as all the other Magpies of the world. It steals anything which glitters, eats the eggs of other birds, and lives off all sorts of prey, grains, fruits and berries. It has one of the most glorious voices of any bird. It is the symbol of South Australia.

The Crested Goura is not an Australian, but lives in New Guinea. It belongs to a group of birds which is well known throughout the entire world — Pigeons — unlikely as that may seem at first sight. If one looks for a moment at the drawing, one can see that the beak, the eye with the dark ring around it, the form of the feet, the wings and the tail are all like those of Pigeons. What confuses one

is the incredible crest of feathers which this variously coloured Pigeon has on its head. When one sees this crest, one inevitably thinks of the more extravagant hats of women of the last century — perhaps gracious, but certainly somewhat difficult to wear.

The Goura has few worries. It is as big as a large fowl, with a brilliant eye and an assured demeanour. It walks solemnly along the ground of the forest where it was born. There are three varieties of Goura, all with crests, but with slightly different markings. All three are covered with a beautiful mantle of blue feathers marked with black and white on the wing.

The Crested Goura lives well in captivity and is found in many zoos around the world. However, this bird is still quite rare. It is little known and its habits of life are uncertain. Its robustness and its ability to eat many sorts of food permit one to hope that soon it will be reproducing under the good conditions which exist in zoos today.

We must remember that often this is the final argument in favour of such institutions. They allow us to preserve rare species and to make sure that they do not disappear, by assuring their reproduction. This method is often successful and we can hope that some day we will be able to re-introduce the survivors into their countries of origin. This should be thought of by all visitors to zoos, who could then better appreciate the efforts of their staffs. However, this is only successful according to how well sick animals are treated. If they are not well, they will never reproduce. These birds are not the only creatures which are being saved in this way. Thus the Orangoutangs of Borneo, and many other animals, are being preserved in zoos.

THE SATIN BOWER-BIRD
(male and female)
(Ptilonorhynchus violaceus)

Here is the Satin Bower-bird, on the left the male, and on the right the female. Because of them, we see that an artistic sense, a good taste in decoration, a love of beauty (and even of extravagance) are not confined only to humans.

These creatures, which are not particularly attractive to look at, are Bower-birds. When their mating season arrives, in order to seduce and keep the female, the male constructs a very special nest in the middle of some large, dry grass. In front of the nest he makes a sort of garden in which he places all sorts of objects which no doubt seem decorative to him. These include feathers, leaves, flowers, mushrooms, snail-shells etc. Stones are also used, and each species has a preference for a different colour—blue, red or green. When all is ready, the male enters the nest (or if one prefers it, the bower) and waits for the visit of the female. If the decor pleases her, if the colours chosen agree with her taste, the

"marriage" takes place. Immediately afterwards the female makes a nest in a tree, where she goes to lay and hatch her young. The male continues to look after and decorate his garden, and wait for visits by other females, who are not long in coming.

We have shown the Satin Bower-bird in the middle of charming his female. Perhaps the best known kind of Bower-bird is McGregor's Bower-bird. Like many members of his strange family, he lives in the mountains of New Guinea, up to an altitude of 300 metres. His bower is made around a bush, to which he has added numerous decorative objects. He uses the same bower for many years, perhaps for all his life. The female's nest is several metres above the ground and contains only one egg.

Coming back to our Satin Bower-bird, which is Australian, we find that its favourite colour is blue. Not only does it collect objects of this colour, but also it makes a paint by beating a suitable plant and using its saliva. Then taking a piece of bark in its beak, it paints this mixture all around its bower. When everything is finished it dances for joy, whistles, cackles and shows the greatest satisfaction.

What does this mean? At the beginning of this section we have said that the Bower-bird has a sense of Art. Perhaps this is taking an unjustifiably anthropo-

morphic view, i.e. we are looking at the animal through human eyes, imagining human emotions, instead of considering what is actually going on. From what we know of birds and their ways of life, it is very important to them that they be able to mark out a territory which is their own. Instead of using noise and songs for this as Robins do, the Bower-bird decorates its territory with objects and colour. The result is the same. Song or marks indicate to rival males that the territory is occupied and they do not penetrate it, but go elsewhere.

While it is impossible to enter into the mind of a bird and understand its very limited thoughts, all the same we can admire the beauty which they show. We can also admire the many ways in which Nature has insured that the various species will propagate, without infertile mis-matings taking place. A female Bower-bird of the sort which collects blue objects is never attracted by red ones, and vice-versa, so there is no mixing of species and no errors.

The Bower-bird is extremely rare in zoos throughout the world. It does not stand up to captivity very well. Perhaps it longs for its own bower, which it built in its native region. The Explorers, Zoologists and Ornithologists who would like to study it have to go to where it lives in Australia and New Guinea. The Bower-bird does not domesticate easily.

1. **THE DIAMOND FIRE-TAIL**
(Emblema guttata)

2. **THE LONG-TAILED FINCH**
(Poephila acuticauda)

3. **THE GOULDIAN FINCH**
(Chloebia gouldiae)

4. **THE RUFOUS-CROWNED EMU-WREN**
(Stipiturus ruficeps)

5. **THE RED-WINGED WREN**
(Malurus elegans)

6. **THE TAWNY FROG-MOUTH**
(Podargus strigoides)

Australia is a country where there are marvellous little birds, very colourful, very cheeky and much sought after by caged-bird lovers. In fact they do not seem very unhappy in cages, and are sometimes sent long journeys in aircraft in this way.

The first three are Finches. These are beautiful little birds which eat grain. They are robust, always covered with feathers, which have very clear and decorative colours. They usually make their nests in the form of a bowl, or a purse with an opening in the side. They are often built in the roof of a shed in a garden or under an old, dead tree. Finches live in little groups and fly about looking for food amongst bushes and tall grass. Often each one is marvellously multi-coloured.

The next two birds carry their tail at a considerable angle, as can be seen from the figures. They flit about in the open country, like little aerial mice.

But let us talk for a moment about the unusual Tawny Frog-mouth. In any country in the world this bird could pass unnoticed, because it is a bird of the dusk, which flies silently, almost invisibly, chasing insects with its large mouth open. Its beak is quite soft, unlike that of most other birds. It is quite short, with hairs and moustaches on it, which help it to catch its prey. By night it chases nocturnal insects; by day it rests on the ground, or on a branch of a tree. It lies along the branch and never perpendicular to it, because it always wants to camouflage itself. It lays rough, speckled eggs, without even a nest to look after them, but they are invisible unless one knows what to look for. Its legs are very small and not of much use for walking. It has a certain difficulty in taking off when it is on the ground. Nevertheless it is often on the ground. In flight, it is often confused with the owl, of which there are many in Australia.

It seems superfluous to say that Tawny Frog-mouths are useful birds. However, it should be repeated, in the hope that this may stop the death of a single one of them. They kill so many insects that the presence of a few pairs of Frog-mouths in an area is a considerable help in this regard. Still some people fear them because of superstition.

As with other similar birds, when flying, they make a bizarre sort of noise, which in Europe has been said to be like the sound of spinning-wheels. This should not disturb people, but it has scared a number. Unhappily one seldom sees them catching insects, since they do it at night. However this is a fascinating sight for those who have the good fortune to see it, as when the light of the moon shows them up.

THE REPTILES

1. THE INSULAR GECKO
(Phyllurus louisiadensis)

2. THE MOUNTAIN DEVIL
(Moloch horridus)

3. THE SHINGLE-BACK SKINK or SLEEPY LIZARD
(Tiliqua rugosa)

4. THE WESTERN BLUE-TONGUED SKINK
(Tiliqua occipitalis)

Here are four reptiles, chosen from among the great number which live in Australia. All four look as if they might be hurtful, but in reality these are quite inoffensive little animals. One need not fear them, because they have no poison and do not attack, but it is unwise as well as unkind to tease those depicted in figures 3 and 4. While their teeth are hard to see, they can give a rather nasty bite.

These are typical Australian animals. The Gecko, like the Geckos of other hot countries, is a hunter of insects and flies, and is capable of climbing anywhere and running up the walls of houses. It is a charming little animal, which moves very quickly and abruptly, and is extremely useful.

The Mountain Devil, which is also called the Devil Lizard or the Hairy Devil, is only 15 to 20 centimetres long. In spite of the frightening spines which cover its yellow and brown scales, it is a charming little animal, quiet, gentle and rather sleepy. It eats small insects and ants, and a number of people keep it as a pet. While the Gecko is often found near cities, the Mountain Devil is found only in the central deserts of Australia. It seems to be a very close relative of a colossal lizard which also had many spines and which was living up to prehistoric times in Australia. Specialists call it *Megalania prisca*. This was an animal more than 6 metres long which may have been responsible for the legends of the Bunyip amongst the Aborigines.

The skinks are short-legged lizards. They love to live in warm regions in the sands of the desert, and seem almost to swim over the sand. They also live in other rocky or wooded regions. Here are two curious specimens; one is called Sleepy Lizard or Stumpy Tail, and has very large, rough scales and a short head. He looks rather like the venomous lizards of Mexico, the Gila Monsters, but he does not have any poison. However, as we mentioned before, one should be careful as he can give a nasty bite. Perhaps this is only fair, if he does this to a hand which annoys him.

The Blue-tongued Skink has another method of

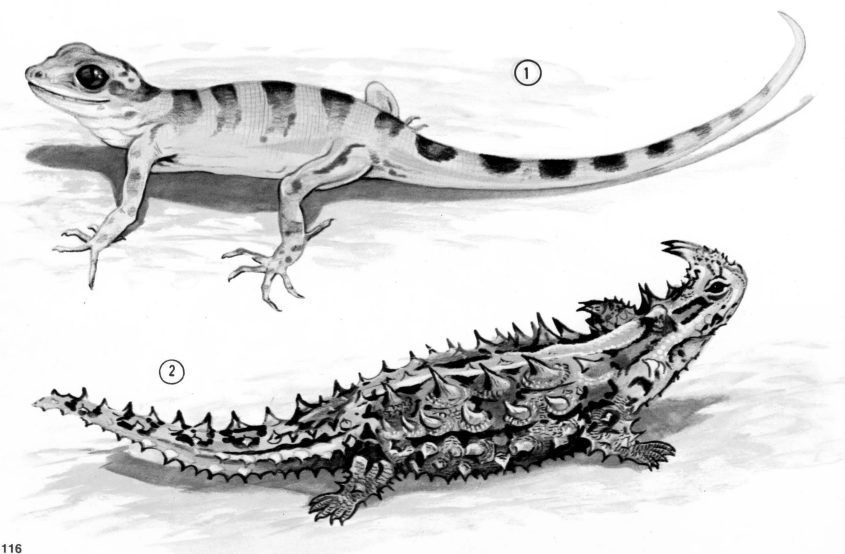

defending himself. He shoots out from his mouth, which opens very widely, a large, thick, blue tongue, which frightens any enemies which might attack. Then the owner of this amazing tongue takes advantage of their surprise, by disappearing in the sand. It has been well said, that of all the innumerable methods by which weakly-defended animals frighten and worry their enemies, this is the most astonishing and most original. The Blue-tongued Skink is found to some extent throughout Australia, where he is sometimes thought to be dangerous. On the contrary, he is quite harmless and destroys many noxious insects. While one can often see Blue-tongued Skinks in zoos, they do not flit out their tongue unless they are frightened, which is a pity because it is an astonishing sight when they do so. Australia is decidedly a country of most unusual surprises.

In spite of what has been said about the Blue-tongued Skinks frightening their enemies, Dingoes and Aborigines are not frightened. Both of these like eating the lizards, and pick them up no matter how often they put out their tongue. The Aborigines like to cook them over the embers of a fire. They chase these lizards with a speed and a patience which is amazing. Nothing stops them and, with the aid of a simple pointed stick, they find the means of survival in the desert, which is the hottest and driest and most hostile in the world. When they are hungry no animal can escape them and they often come back to camp with half a dozen different lizards suspended from their belt.

1. **THE DIAMOND PYTHON**
   (Morelia spilotes variegata)

2. **THE DEATH ADDER**
   (Acanthophis antarcticus)

3. **THE TIGER SNAKE**
   (Notechis scutatus)

Australia is equally rich in snakes as it is in other animals. It is said that there are more poisonous snakes on this continent than anywhere else. This is perhaps an exaggeration, but it is certainly true that there are many of them. Here are three typical specimens of this widely distributed type of animal, which both fascinates and frightens one at the same time.

The Diamond Python in figure 1 can be up to 7 metres long. One should remember that no snake, neither the Indian Python, nor the Anaconda of the Amazon has been measured at more than 9 metres. The Great Australian Diamond Python also lives in New Guinea. On the continent of Australia, it is found in northern Queensland, in rocky regions, where it catches Wallabies which it loves to eat. Like all the great snakes, the Diamond Python kills its victims by squeezing them between rings of its body until their heart stops beating. Contrary to legends, it does not break the bones of its panting prey, but suffocates them more or less quickly, before swallowing them head first. It can swallow even a large dog or sheep. It has not been known to attack men unless provoked, pursued or molested.

Much more dangerous are the venomous species shown in figures 2 and 3. The Death Adder is a short, thick animal, with a very big head containing enormous poison glands, leading into long, hollow, curved fangs which inject the poison. Its body takes the main colour of the region where the animal lives and ends in a small, bizarre tail. This it uses as a lure for birds, which come to pick at what they think is a worm. Because of its disinclination to move when approached, it is sometimes called "Deaf Adder". The Death Adder can best be compared with the deadly Vipers of Africa, which are similar in size and are similarly dangerous.

The Tiger Snake (figure 3) is an equally dangerous beast. It is more rapid and more lively than the Death Adder. It is more like a Cobra than a Viper and it is said that a man bitten by this animal has less than an hour to live, unless he gets help at once. Unfortunately it is hard to help oneself when one has been bitten by a snake, so the large distances between towns of Inland Australia makes for an even more dangerous situation. Therefore it is best always to be careful when in regions where there live these snakes, or indeed any others.

THE TUATARA
(Sphenodon punctatus)

The Tuatara lives in New Zealand. With the coming of Europeans, the result has been that dogs, cats, pigs and men chase him and he has become rare except on certain islands. Great efforts are now being made to ensure his survival. He is a big, bright-green lizard, which can be 60 centimetres long, and lives in holes which he digs in the ground under the vegetation. He has a line of soft spines along his back which cease at the start of his tail.

The *Sphenodon* is a true living fossil, without doubt the most remarkable in the entire world. He is the last representative of a line of animals which dominated the world before the coming of the mammals. In the skin on top of his head, just behind his visible external eyes, is an internal third eye, which is called the pineal eye. The pineal gland is possessed by both reptiles and mammals, and is hidden in the cranium. It does not reach the outside. Other reptiles also have, beneath the bones of their skull, the vestiges not only of the gland but also of a pineal eye. However, the *Sphenodon* has this third eye, opening on the skin.

Not only is the third eye visible, but it is functional, because the animal responds to any light which shines on it. This is one of the principal reasons for the interest which the Tuatara creates, and for the protection which it is given. Specimens sent to zoos in other parts of the world live very well in captivity, but now the Government of New Zealand keeps them only in New Zealand and refuses to export them. We should add that it is a peaceful animal, placid and calm, which likes to live tranquilly, warming itself in the sun, while eating insects from time to time.

Australia has Goannas. These are giant lizards, which have their heads attached, not to their bodies as do most other lizards, but on the end of a long neck. The most widely known are the lizards shown here, which may be up to 1.80 metres long. This is not as long as the giant of the species, the Komodo Dragon.

COMBAT OF TREE MONITORS
(Varanus varius)

Goannas are lively, rapid animals, which can climb trees, swim, run quickly, bite, and defend themselves with extremely painful blows of the tail. They eat anything and are omnivorous carnivores. Thus they eat birds, eggs, insects and even little mammals, because of their size and force. To fight each other they raise themselves on their hind-feet like the two shown here. They look like some ancient great lizards from prehistoric times.

Zoologists who have studied Australia say that there is another species, which is very much greater (3 metres), named *Varanus giganteus*. This is supposed to exist in the centre and in the north-west of the country.

But the formal proofs of its existence have not been given. The Naturalists of the world eagerly await such proofs, because new animals are very seldom found and their discovery would make a great impact.

It is certain that Australia, New Guinea and perhaps also the centre of South America are, with the bottoms of the oceans, the last regions of the world where new zoological discoveries are likely to be made. Apart from the animals that live in the depths of the sea, it is only those which are rare or of very small size which are likely still to be awaiting discovery, The possibility of the existence of the marsupial Tiger and the giant Varan gives hope that young Zoologists may still discover major animal species.

We cannot leave the animals of Australia without showing two which live on the borders of rivers and oceans. They are extremely dangerous because one of them is one of the most terrible canivores of the world while the other is one of the most poisonous reptiles.

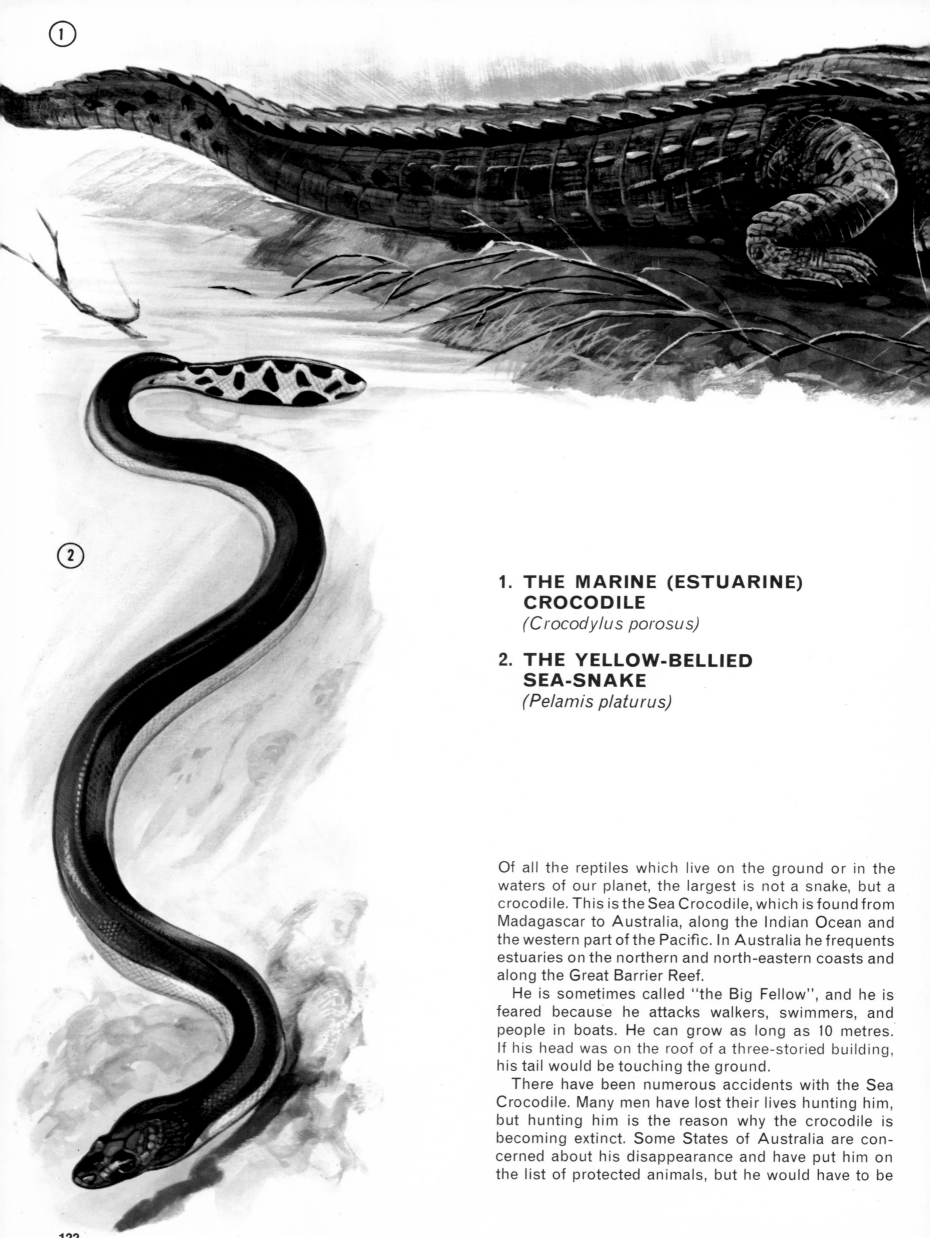

1. THE MARINE (ESTUARINE) CROCODILE
(Crocodylus porosus)

2. THE YELLOW-BELLIED SEA-SNAKE
(Pelamis platurus)

Of all the reptiles which live on the ground or in the waters of our planet, the largest is not a snake, but a crocodile. This is the Sea Crocodile, which is found from Madagascar to Australia, along the Indian Ocean and the western part of the Pacific. In Australia he frequents estuaries on the northern and north-eastern coasts and along the Great Barrier Reef.

He is sometimes called "the Big Fellow", and he is feared because he attacks walkers, swimmers, and people in boats. He can grow as long as 10 metres. If his head was on the roof of a three-storied building, his tail would be touching the ground.

There have been numerous accidents with the Sea Crocodile. Many men have lost their lives hunting him, but hunting him is the reason why the crocodile is becoming extinct. Some States of Australia are concerned about his disappearance and have put him on the list of protected animals, but he would have to be

protected everywhere for him to have a good hope of survival.

They reproduce by means of eggs. These are laid on land and the number in a clutch may vary from twenty to ninety, depending on the size of the mother. The babies are about 18 centimetres in length at birth. Actually it is the young, measuring less than a metre long, which are most often killed to make crocodile-skin bags and shoes for women. The skins of the large crocodiles are not usable, as they are too tough and horny.

In Queensland, in the mangrove swamps, there live many of these crocodiles, which are rarely attacked there. They get bigger and bigger as they get older. Their eyes have two pupils, one of which lets them see under the water, and the other above the water. Their nostrils are located on the tip of the snout and are equipped with a tube and a valve. The tube extends well down the throat, allowing the animal to breathe freely when partially submerged; the valve enables them to feed under water. They love to lie idly in the water near the edge of a river, with only their eyes and nostrils showing. Any unwary bird or animal that comes within reach of the powerful jaws is seized, dragged under water and eaten. They swim very strongly, their powerful tails serving as an oar. They can come up the side of a boat and take someone, wounding or killing them. They are very dangerous animals.

The last Australian animal which we are showing is the Sea-Snake, which is common amongst the warm seas of the Pacific. There are many varieties, which are specially adapted to living in their environment. They are all characterised by a flat tail with which they swim. They are up to 2.5 metres long and can inject their poison by means of two fangs in their upper jaws. The venom is extremely dangerous and is usually fatal to men. It is contained in two glands placed below the articulation of the jaws. Sea-Snakes have nostrils on the tip of the snout, with a valve to enable them to feed under water. They must come to the surface to breathe from time to time, and they lay their eggs in the sand of the beach. They do not usually attack bathers, but are dangerous when they are found in nets or are cornered. They eat fish and reptiles.

There are also venomous shells in this region, such as Cone Shells, which have one of the most deadly poisons in the world.

We now come to the end of our survey of the animals of Australia. Many are beautiful, many are interesting, many are useful. As civilization spreads more and more, and as more and more of the country is used to produce food, or for mining minerals, we must be careful that we do not lose these animals. While we must use the country in order to live, we must also take care that species do not vanish completely. Indeed, it is better that they should become more numerous in some parts. In this way life will be more interesting and more varied not only for us but also for those who come after us.

Unfortunately, not only the correct region but sometimes quite a lot of land is needed in order that a species may survive. Sometimes this must be good land, which people would naturally like to use for other purposes, such as for rearing sheep and cattle. But some good land must be used for our priceless heritage of wild animals. We are only the custodians of our country: it is ours to use and enjoy, but not to squander. We should aim to deliver it to future generations in a better condition for our having lived in it.

INDEX